A Short History of Vertebrate Palaeontology

Frontispiece: 'The trumpet of scientific judgment has sounded. They have risen from the dead, and the naturalist classifies them': a late nineteenth-century conception of vertebrate palaeontology (Flammarion, 1886).

A SHORT HISTORY OF
VERTEBRATE PALAEONTOLOGY

ERIC BUFFETAUT

CROOM HELM
London • Sydney • Wolfeboro, New Hampshire

© 1987 Eric Buffetaut
Croom Helm Ltd, Provident House, Burrell Row,
Beckenham, Kent BR3 1AT
Croom Helm Australia, 44–50 Waterloo Road,
North Ryde, 2113, New South Wales

British Library Cataloguing in Publication Data

Buffetaut, Eric
    A short history of vertebrate palaeontology
    1. Vertebrates, Fossil   2. Palaeontology—
    History
    I. Title
    366′.09   QE841

    ISBN 0-7099-3962-0

Croom Helm, 27 South Main Street,
Wolfeboro, New Hampshire 03894-2069

Library of Congress Cataloging in Publication Data
applied for.

Phototypeset by Sunrise Setting, Torquay, Devon
Printed and bound in Great Britain

Contents

List of Illustrations

Frontispiece: 'The trumpet of scientific judgment has sounded. They have risen from the dead, and the naturalist classifies them': a late nineteenth-century conception of vertebrate palaeontology (Flammarion, 1886)

Fig. 1: The influence of the unicorn myth on early reconstructions of vertebrate fossils: Otto von Guericke's drawing of a fossil skeleton, based on mammoth bones from Quedlinburg (after Leibniz, 1749)

Fig. 2: Scheuchzer's *Homo diluvii testis*: a giant salamander from the Miocene of Oeningen

Fig. 3: Exotic animals in the ground of western European countries: reindeer antlers from Etampes, illustrations by Guettard (1768)

Fig. 4: The 'American Incognitum': a mastodon tooth from North America, illustrated by Buffon (1778)

Fig. 5: A famous episode of eighteenth-century palaeontology: the discovery of the 'Great Animal of Maastricht' in the underground quarries of Saint Peter's Mountain (Faujas de Saint-Fond, 1799)

Fig. 6: A portrait of Georges Cuvier, from the last, posthumous, edition of his *Recherches sur les ossemens fossiles* (1834–6)

Fig. 7: The 'animal from Paraguay', actually a giant ground sloth from the Pleistocene of Argentina, which was mounted in Madrid by Bru, and described by Cuvier as *Megatherium*

Fig. 8: The earliest scientific reconstructions of fossil vertebrates: Eocene mammals from the Montmartre gypsum, drawn by Laurillard under Cuvier's supervision (from Cuvier, 1834–6)

Fig. 9: The 'age of reptiles' as envisioned by Thomas Hawkins (1840): a fight between an ichthyosaur and two plesiosaurs on a Jurassic seashore

For Haiyan

Foreword

Vertebrate palaeontology can no longer be called a young science. It became formalised at the very end of the eighteenth century when the former existence of now extinct species was accepted by a growing number of scientists, and when the methodological bases for the reconstruction and identification of fossil skeletons were established. Thus, vertebrate palaeontology, like palaeontology as a whole, is definitely older than other disciplines also concerned with the evolution of life, such as genetics or molecular biology. This means, among other things, that the history of vertebrate palaeontology, if told in detail, could fill many volumes. This book is not meant as a complete and exhaustive history of vertebrate palaeontology. Rather, its aim is to relate the major events and phases in the study of fossil vertebrates, from its pre-scientific beginnings to, roughly, the First World War, with a few comments added on more recent developments. Despite the close links between vertebrate palaeontology and these subjects, this is not a history of geology or of evolutionary theory, although mention is often made of episodes in the history of these fields. That the study of fossil vertebrates should be considered as a separate entity within palaeontology as a whole is justifed by the historical development of the discipline: from the beginning, it had closer links with biological sciences such as anatomy and zoology than invertebrate palaeontology, which was often practised more as a branch of stratigraphy than anything else. This of course does not mean that there were no biologically minded invertebrate palaeontologists, but their role in the general development of their science was smaller than that of their counterparts working on vertebrates. Another characteristic of vertebrate palaeontology is the importance of individual fossils and fossil localities. Many important forms of fossil vertebrates are known only by a very few specimens (*Archaeopteryx* and *Compsognathus* immediately come to mind, but many less famous examples could be mentioned). Moreover, although thousands of fossil vertebrate localities are known, some of them are of particular and lasting importance, and their discovery and scientific exploitation often marks an important episode in the history of vertebrate palaeontology. Both features are of lesser importance in the development of invertebrate palaeontology, because fossil invertebrates are generally much more abundant and widespread than vertebrate remains.

In writing this concise history, I have concentrated on some especially important events which I think can be considered as turning points in the evolution of vertebrate palaeontology. A first essential step was the general recognition of the organic nature of fossils at the beginning of the eighteenth century. After this came the acceptance of the phenomenon of extinction, although this did not become general until the early years of the nineteenth century. The discovery of a succession of extinct faunas, mainly under the influence of Blumenbach and Cuvier, was also a very important conceptual development, while Cuvier's methodology formed the basis of the subsequent progress of the discipline. The next major step was, of course, the acceptance of evolution, which became general only after Darwin had provided a reasonably convincing mechanism. Post-Darwinian vertebrate palaeontology was mainly concerned with the accumulation of fossil evidence and its interpretation in evolutionary terms. The late nineteenth century and early twentieth century also saw the spread of research on fossil vertebrates to all parts of the world, with its consequences on the biogeographical interpretation of vertebrate evolution.

Although there are relatively few books on the general history of vertebrate palaeontology, particular episodes have received much attention. There are, for instance, several competent and well-written accounts of the great palaeontological discoveries made in the American West in the nineteenth century, and of the famous 'fossil war' between Cope and Marsh. Consequently, I have not attempted to give detailed accounts of such famous episodes, but have concentrated instead on less well-known developments, or on those which are known to specialists but are not usually described in much detail in general histories of palaeontology. Although the search for fossil man occupied a large number of prominent vertebrate palaeontologists during the nineteenth century, I have chosen not to include the story of palaeoanthropology in this book for the simple reason that it deserves (and has received) independent treatment at book length: space simply was not available for this.

The reader will find quite a number of anecdotes in this book. I hope they will not be considered as trivial. My opinion is that they are often revealing of the intellectual and social context in which palaeontological research was conducted at various periods of its history, and that they can provide valuable insights into the psychology of some eminent or little-known palaeontologists.

Finally, I expect that some palaeontologists will be disappointed not to find a mention of certain discoveries or interpretations which they consider as important for the history of research in their special field. Unavoidably, choices had to be made for reasons of space, and they undoubtedly reflect to some extent the personal preoccupations and tastes of the author.

Eric Buffetaut
Paris

NOTE

For easier reading, all quotations from works in foreign languages have been translated into English. Unless indicated otherwise, the translation is by the author.

1

Pre-scientific Notions About Fossil Vertebrates

The earliest-known 'fossil collection' was found in a Middle Palaeolithic level in a cave at Arcy-sur-Cure, in Burgundy: it consisted of a gastropod and a coral, and the collector was apparently a representative of Neanderthal man (Leroi-Gourhan, 1955). That invertebrate fossils should have been noticed and collected by man before vertebrate remains is hardly surprising, for the simple reason that fossil invertebrates are usually much more abundant than vertebrate bones in sedimentary rocks, and more conspicuous because of their easily recognisable shapes; vertebrate fossils, when exposed by erosion, are often much fragmented and less easy to identify than, for instance, fossil shells. Nevertheless, some vertebrate fossils with a peculiar shape are sufficiently common to have attracted the attention of prehistoric man. Fossil fish teeth, with their attractive shiny dark enamel, are fairly abundant in some localities, and their shape can be distinctive, as in the case of the blade-like teeth of sharks or the button-shaped teeth of *Lepidotus* and pycnodonts. Such teeth have been found in many archaeological sites, where they had been brought by prehistoric men. Oakley (1975) has listed a number of instances of fossil shark teeth from late Palaeolithic sites in southern France, some of them a considerable distance from the nearest fossil locality. These teeth were probably considered as valuable objects, and there is no doubt that they were sometimes used for decorative purposes, as some of them are perforated and were apparently parts of necklaces or pendants. According to their finder, Count Henri Bégouen, two *Isurus* teeth found in an Aurignacian layer in the Tuc d'Audoubert cave (in the French Pyrenees) may have been collected from Miocene rocks some 150 km (93 miles) distant from the cave; they are provided with 'a pair of biconical holes through the roots for securing suspension in a well-balanced position' (Oakley, 1975, p. 16). Southern France is not the only region where late Palaeolithic man collected fossil shark teeth: such a fossil has been found in a Magdalenian layer in the Petersfels cave in southern Germany (Peters, 1930). More unusual types of fossil fish teeth were sometimes found by Palaeolithic man: Oakley (1975) mentions a crushing tooth of the hybodont shark *Asteracanthus* from the famous Grimaldi caves on the Riviera, as well as a toothplate of the Triassic lungfish *Ceratodus* from the Kesserloch cave at Thayngen in Switzerland.

In later archaeological sites, fossil shark teeth have also been found; Oakley (1975) lists several cases from Neolithic, Predynastic and Dynastic times in Egypt. In Dynastic times such fossils were apparently used as amulets, and sometimes set in metal. This was also done by the Etruscans, as shown by a tooth mounted in filigree, now in the Ashmolean Museum, Oxford, illustrated by Oakley (1975, pl.Id). Teeth of the huge shark *Carcharodon* were collected from Miocene beds on Malta by Chalcolithic and Bronze Age men, and it seems that the serrated edges of these teeth were used for making impressions on clay pots before they were fired (Oakley, 1975).

How prehistoric men interpreted the vertebrate fossils they sometimes found and used for decorative (and possibly magical) purposes, we shall never know. It took a long time for fossil shark teeth to be recognised for what they were, and rather peculiar conceptions developed about them during Antiquity and the Middle Ages. They were usually known as 'glossopetrae' or 'tongue-stones', because of their supposed resemblance to the forked tongues of snakes, and one popular interpretation was that they were the petrified tongues of snakes. The *Carcharodon* teeth from Malta mentioned earlier, gave rise to the legend according to which Saint Paul, shipwrecked on this island on his way to Rome, was bitten by an adder, which prompted him to curse all snakes on the island, with the result that their forked tongues were turned into stone (Bassett, 1982). The Maltese glossopetrae, sometimes referred to as 'St Paul's tongues', were famous until relatively recent times, and are mentioned in many old works on 'petrifactions'.

The interpretation of fossil shark teeth as petrified snake tongues led to the belief that they could be used as a protection against snake bite and poison. This was first based on sympathetic magic, and glossopetrae were worn as amulets. Later, it was supposed that they possessed properties enabling them to absorb or neutralise poison, and they were dipped in beverages which might contain poison (Oakley, 1975). From the Middle Ages until the eighteenth century, elaborate tree-like structures, sometimes made of precious metals, were used to suspend glossopetrae and keep them ready for use on dining tables. A few such 'languiers' or 'Natterzungenbäume' have been preserved till the present day, and some are illustrated by Oakley (1975).

Fossil shark teeth also were given other, no less fanciful, interpretations. The Roman naturalist Pliny the Elder (23–79 AD), for instance, thought that the glossopetrae resembled a man's tongue. He believed that they fell from heaven during eclipses of the moon and were 'thought by the magitians to be very necessary for pandors and those that court fair women' (quoted by Edwards, 1976). The supposedly celestial origin of the glossopetrae was accepted by many and was linked with the frequent confusion between shark teeth and prehistoric stone axes, known as cerauniae, the real nature of which was not recognised until Mercati's work in the sixteenth century. Both the cerauniae and the glossopetrae had a pointed, more or less triangular, shape, which accounts for the confusion. The popular idea — also accepted by some scholars — was that the cerauniae were 'thunderstones', which fell from heaven during thunderstorms and were found in the ground in places which had been struck by lightning (Laming-Emperaire, 1964). It was not until the seventeenth century that the real nature of the glossopetrae was finally recognised and accepted (see Chapter 2).

Other kinds of fossil fish teeth received different interpretations. The rounded, button-like teeth of the Jurassic fish *Lepidotus* were known as bufonites, or toad-stones, and thought to be jewels formed in the heads of toads. Toad-stones were supposedly endowed with medicinal or magical properties (Bassett, 1982).

Fossil vertebrates played a very limited part in the 'geological' speculations of some Greek authors such as Xenophanes (born in 617 BC), or Herodotus (484–425 BC), who had recognised that some fossils were remains of marine organisms, and could be used to demonstrate a former greater extension of the sea; their speculations were mainly based on fossil sea-shells. Neverthe-

3

less, fossil bones did attract the attention of the ancient Greeks, and gave rise to stories and legends. An interesting case is that of Miocene mammal bones from the island of Samos, in the Aegean, about which there were local myths. According to Aelianus (quoted by Solounias, 1981, p. 18).

> Ephorion says in his memoirs that in ancient times Samos became uninhabited because of the appearance of great fierce beasts which prevented the people from travelling. These beasts were called Neades, and they could fracture the earth with the sound of their voices alone. The people of Samos had a proverb: to scream louder than the Neades. The author says that even now one can see huge bones of these beasts.

Plutarch's explanation was different: the bones, which could still be seen on Samos, were those of the Amazons who were slaughtered there by Dionysus (Forsyth Major, 1891; Solounias, 1981). In the 1880s the British palaeontologist Forsyth Major rediscovered the legends, and came to the conclusion that they were probably based on fact. His subsequent search for fossil vertebrates on Samos was rewarded by the discovery of Miocene localities, which he excavated in 1887 and 1889 (Forsyth Major, 1891; Solounias, 1981). This marked the beginning of the scientific exploitation of these now famous vertebrate localities, which continues to the present time.

On the island of Kos (or Cos), about 100 km (62 miles) southwest of Samos, direct evidence of the interest in fossil vertebrates among the ancient Greeks was found by the American palaeontologist Barnum Brown in the 1920s, when he discovered a piece of a fossil elephant tooth in the ruins of the Asklepeion, a famous medical school where Hippocrates is supposed to have studied in the fifth century BC. Whether Hippocrates himself actually handled and discussed the specimen, as suggested by Brown (1926), can of course never be demonstrated, but there is no doubt that the tooth had been considered sufficiently valuable to be carried from one of the vertebrate localities on Kos to the Asklepeion by the ancient Greeks.

Bones of large fossil mammals, notably proboscideans, are among the most noticeable of vertebrate fossils. They occur relatively often in Pleistocene deposits, and many early finds attracted much attention. Some of the reports date back to classical Antiquity. Empedocles (492–432 BC) had already reported finds of large bones in Sicily and considered them the remains of a race of giants (Abel, 1914). Pliny mentioned fossil ivory found in the ground, and according to Suetonius, the Emperor Augustus owned a collection of huge bones which had been found on the island of Capri; they were also said to be those of giants (Archiac, 1864).

During the Middle Ages and the Renaissance, and until the end of the seventeenth century, bones of large fossil mammals were generally believed to be the remains of human giants or mythical monsters, and the early history of vertebrate palaeontology (if it may so be called at such an early stage) is largely a story of giants and dragons. Comparative anatomy did not really begin to develop until the late seventeenth century, and until then the most sensible explanation for huge bones found in the ground was that they belonged to those giants whose existence was attested by Scripture as well as by mythology. Genesis was unequivocal about this: 'there were giants in those days'. It is not surprising, therefore, that old chronicles should frequently mention discoveries of the bones of giants; such reports excited the imagination of laymen and scholars alike, and 'gigantology' became a minor branch of specialised knowledge.

The giants of classical mythology provided an easy explanation for some finds. Remains of Pleistocene dwarf elephants found in a Sicilian cave near Trapani in the fourteenth century were identified as those of the Cyclops Polyphemus by the Italian writer Giovanni Boccaccio (1313–75), the famous author of the *Decameron*. According to the Austrian palaeontologist Othenio Abel (1914), fossil elephant skulls may even have led to the very idea of one-eyed giants, their large median nasal opening having been mistaken for a single orbit. Abel supposed that Greek sailors found such elephant skulls in Sicilian caves, and that such discoveries gave rise the Homeric story of Ulysses and Polyphemus. This attractive hypothesis can hardly be tested, but there is no doubt that belief in giants was considerably strengthened by finds of elephant or mammoth bones. In some instances, these gigantic bones were thought to be those of Christian saints: in Valencia, for instance, a mammoth molar was kept as a relic of Saint Christopher (Pfizenmayer, 1926), and in Munich a dorsal vertebra of an elephant was venerated as belonging to this saint (Abel, 1939a). More frequently, they were attributed to unidentified giants, but they always attracted much attention.

Giant bones were often kept in churches, castles or public buildings. In 1443 the femur of a mammoth was found in Vienna when the foundations of a tower of Saint Stephen's cathedral were being built. The bone was adorned with Emperor Friedrich III's motto 'A.E.I.O.U.' (*Austriae est imperare orbi universo*) and was kept for a long time in the cathedral (it was still there in 1729), before being transferred to the University of Vienna (Abel, 1914; Pfizenmayer, 1926).

In 1577 large bones were found near the abbey of Renden in Switzerland, between the roots of an old oak tree which had been uprooted by a storm. The city council of Luzern interpreted the bones as those of an angel fallen from heaven, but a physician from Basel, Felix Plater, identified them as those of a 19-foot-tall (5.75m) giant. A picture of this 'giant or wild man of Renden' was painted on the tower of the city hall in Luzern, accompanied by an explanation in verse, but the bones themselves were given a Christian burial (Pfizenmayer, 1926; Abel, 1939a). Similarly, a mammoth tusk was displayed in St Michael's church in Schwäbisch Hall, in south-western Germany, accompanied by the following written commentary (Pfizenmayer, 1926):

In the year sixteen hundred and five
On February thirteenth I was found
Near Neubronn in the region of Hall
Tell me, dear, to which species I may belong

Many additional examples from various parts of central Europe (Hungary, Silesia, etc.) have been given by Abel (1939a).

At the beginning of the seventeenth century, a discovery of proboscidean remains in south-eastern France caused a bitter and protracted controversy among scholars. This famous case of the 'giant Theutobochus' has recently been reviewed by Ginsburg (1984), and it is worth recounting in some detail, as it is a good illustration of the kinds of interpretations which were fashionable at the time. In January 1613 workmen were digging in a sandpit near the castle of Chaumont, or Langon, four leagues from the city of Romans, in Dauphiné (a province of south-eastern France), when they came across some large bones. These were brought to the local squire, Marquis Nicolas de Langon, who consulted some savants from the University of Montpellier, who thought the bones were human. The Duke of Lesdiguières, governor of Dauphiné, borrowed some of the bones and sent them to 'experts' in Grenoble, who also pronounced them to be human. At this point, a certain Pierre Mazurier, a barber-surgeon in the nearby town of Beaurepaire, comes into the story. He borrowed some of the giant bones from the Marquis de Langon, and started to travel from town to town, exhibiting them for money. A small booklet written by a certain Jacques Tissot (about whom little is known) was sold to the interested public. Its title is worth translating here, as it nicely summarises Mazurier's interpretation of the bones:

> Truthful discourse of the life, death and bones of the giant Theutobochus, king of the Teutons, the Cimbri and the Ambrones, who was defeated 105 years before the coming of our Lord Jesus Christ. With his army, to the number of four hundred thousand fighters, he was defeated by Marius, the Roman consul, and was buried near the castle called Chaumon, and nowadays Langon, near the city of Romans, in Dauphiné. There, his grave was found, 30 feet [9m] in length, on which was written his name in Roman letters, and the bones found therein are more than 25 feet [7.5m] in length, with one of his teeth weighing 11 pounds [5kg] as you will really see in this city, all being monstrous in height as well as in size.

The Teutons, Cimbri and Ambrones were Germanic tribes which wandered across western Europe at the end of the second century BC, thus devastating parts of Gaul and Spain. After defeating several Roman armies sent against them, they finally suffered a severe setback in south-eastern France when they were defeated by the Roman consul Marius in 102 BC. The Cimbri, however, managed to cross the Alps into northern Italy, where what remained of their army was finally destroyed by Marius in 101 BC.

The supposedly historical bones of Theutobochus were eventually brought to Paris by Mazurier and his associate Chenevier. It seems that there was an agreement between Nicolas de Langon and Mazurier, according to which the latter was to return the bones within eighteen months, unless the king wanted to keep them. King Louis XIII, then aged eleven, did express some interest, for the bones were taken to Fontaine-bleau, where he was staying, and exhibited in the queen mother's room. The king's intendant for medals and antiques, who had taken charge of the specimens, even wrote to Nicolas de Langon in August 1613 to ask him to send the remaining bones, together with the inscribed stone bearing Theutobochus's name, the remaining bricks of the grave, and medals which were purported to have been found with the bones. Nothing was sent, however, and the king did not keep the bones. Instead, they were shown again to the public by Mazurier, who exhibited them in northern France and Flanders.

Apparently, doubts were expressed as to the real nature of the bones from Langon, which prompted an answer by Nicolas Habicot, a surgeon and anatomist, in the form of a booklet entitled *Gigantostéologie, ou discours des os d'un géant* (Gigantosteology, or discourse on the bones of a giant). In it, Habicot defended Mazurier's interpretation against those who thought the specimens were not bones at all, or were those of an elephant or a whale. He reminded his readers that the existence of giants was proved by the Bible as well as by mythology and history, and he compared the bones exhibited by Mazurier with those of a man. His conclusion was that they had indeed belonged to a human giant, and that this giant must have been Theutobochus, as shown by geographical and historical evidence drawn from Roman historians.

Nicolas Claude Peiresc, a lawyer from Aix, did not agree with this identification; he pointed out several inconsistencies and improbabilities in Mazurier's account of the discovery, and did not accept that the bones were those of Theutobochus. The controversy became much more bitter, however, at the end of 1613, when Riolan, a famous professor of anatomy and botany at the Royal College of Medicine, published an anonymous pamphlet entitled *Gigantomachie pour répondre à la gigantosteologie* (Gigantomachy, to answer the gigantosteology). This was mainly a violent attack on Habicot's ability as an anatomist, and its chief purpose, apparently, was to ridicule the entire corporation of surgeons. Riolan's interpretation of the bones was that they may have been those of an elephant.

This started a protracted controversy between the physicians and the barber-surgeons, in which the protagonists quickly resorted to the basest invective. Riolan, having been personally attacked, fought back in 1614 with a second anonymous pamphlet in which he intended to expose what he called 'the hoax of the supposed human bones falsely attributed to king

Theutobochus'. He considered that the whole story of the grave, the medals and the inscriptions had simply been invented by Mazurier. He also no longer thought that the bones were those of an elephant, but supposed that they were bone-like stones engendered by the earth (this was the famous idea of the formation of fossils by a mysterious *vis plastica* which had already been applied to fossil elephant tusks by the Italian anatomist Fallopio in the sixteenth century).

In 1615 Riolan's pamphlet was answered in a *Discours apologétique* by an anonymous Parisian surgeon, who attacked both Habicot, for risking his reputation by publishing inept conclusions on giants, and Riolan, for doubting the veracity of biblical and historical records of giants. The main point of the pamphlet may have been to defend the surgeons and to ask for more benevolence from the Faculty of Medicine.

Habicot answered by quoting the Bible and various historians again, and by defending the surgeons. Riolan's reaction was a ferocious personal attack on Habicot and the author of the *Discours apologétique*, followed, in 1618, by a 168-page book entitled *Gigantologie. Histoire de la grandeur des géants* (Gigantology. History of the size of giants). In it, he claimed that human giants had never existed, and that the bones from Dauphiné were either those of a large animal such as an elephant, or had been formed by natural processes inside the earth.

The controversy went on for some time, until Habicot published his *Antigigantologie ou contre-discours sur la nature des géants* (Antigigantology, or counter-discourse on the nature of giants), also in 1618. In it, he quoted a letter from Nicolas de Langon in which the latter explained why the official certificate written at the time of the discovery had never been sent to Paris (the certificate in question was not published until the eighteenth century). He still accepted the conclusions of the experts from Montpellier and Grenoble who thought the bones were human.

The controversy seems to have died down after this publication, but there was an unexpected sequel in the 1830s when mastodon bones were found in an old theatre in Bordeaux, and sent to the Natural History Museum in Paris as those of Theutobochus. Blainville, who was the head of the Comparative Anatomy Department of the Museum, first agreed that they were indeed the bones found at Langon and subsequently exhibited by Mazurier (Blainville, 1835). Two years later, however, an heir of the Langon family sent him some written documents as well as a fossil tooth and two silver medals, said to have been found together. Blainville (1837) then changed his mind and decided that the fossils from Bordeaux were not the original material which had caused the Theutobochus controversy, and identified the tooth from Langon as that of a large rhinoceros.

This tooth has recently been re-examined by Ginsburg (1984), who has identified it as belonging to the proboscidean *Deinotherium giganteum*. There seems to be little doubt that what the workmen found at Langon in 1613 was a *Deinotherium* skeleton. Additional fossils from the same locality indicate a late Miocene age (Ginsburg, 1984).

In most accounts of this famous find and of the ensuing controversy, Mazurier has been denounced as a hoaxer and charlatan, mainly because of his fantastic description of the grave and its inscription. Ginsburg (1984) has tried to exonerate him from these accusations, by claiming that sedimentary structures may have been mistaken for a brick grave and an inscription, and that the medals were accidentally mixed with the Miocene bones. More than 350 years after the events, it is of course difficult to reach a definitive conclusion, but it seems obvious that Mazurier's interpretation can hardly have been based only on observations made at the time of the discovery. The whole Theutobochus story sounds like a scholarly invention more than anything else (it is very unlikely that many people in Dauphiné in 1613 were familiar with the rather obscure story of Marius' victory over the Cimbri, and that the workmen who found the fossils spontaneously interpreted vague sedimentary figures as Theutobochus' name written in Roman letters!) Whether the story was honestly meant to explain the occurrence of large bones in that part of Dauphiné, or was essentially a device to exploit public credulity, is difficult to decide, but it seems unlikely that Mazurier was totally innocent in this affair.

Finds of large vertebrate remains did not always result in such heated controversies, but they usually attracted much attention, even under circumstances not really favourable to scientific inquiry. Thus in 1645, during the Thirty Years War, Swedish soldiers occupying the Austrian city of Krems found several enormous 'giant' skeletons when digging a trench. Some of the remains were sent to various places, including a Jesuit church in

Krems. The geographer Merian gave a picture of one of the teeth in his *Theatrum Europaeum*, which clearly showed that the skeletons found by the Swedes were those of mammoths. The tooth illustrated by Merian was eventually rediscovered in a Benedictine monastery near Krems by Othenio Abel in 1911 (Abel, 1939a).

Although the 'giant hypothesis' was still accepted by many authors in the second half of the seventeenth century, alternative explanations were also put forward to account for the presence of enormous bones in the ground. Thus, in his famous *Mundus subterraneus* (1678), the eclectic Jesuit Athanasius Kircher gave a long list of finds of giant bones, but did not accept some of the more extravagant size estimates. According to Kircher, the 300 feet (91m) attributed to the giant Polyphemus by Boccaccio on the basis of bones from a Sicilian cave were an exaggeration; a maximum height of 30 feet (9m) seemed more likely! Kircher also mentioned the possibility that some of the bones found in Sicily may have been those of elephants brought there for military purposes (an explanation which was to become popular when the animal nature of large fossil bones at last became generally accepted). However, he also thought that some bones and teeth could form spontaneously within the earth.

Most stories of giant bones were based on finds of large Pleistocene or Neogene mammals, such as elephants, mammoths, mastodons and dinotheres. One report, however, was based on a bone of a Mesozoic reptile, and it is worth mentioning as the first 'scientific' description of a dinosaur bone (Halstead, 1970; Buffetaut, 1980). In 1676, the Reverend Robert Plot (1640–96), first keeper of the Ashmolean Museum, published his *Natural History of Oxford-shire being an essay toward the natural history of England*. This was one of the early attempts at describing with some accuracy the natural curiosities of a well-defined region, and fossils occupy an important part in this book. Plot generally did not accept the organic origin of most fossils, and followed the opinion of the physician and naturalist Martin Lister (1638–1711), who believed that fossils were produced by a 'plastic virtue' latent in the earth. This explanation, however, did not apply to a large bone 'dug out of a quarry in the parish of Cornwell', which had 'exactly the figure of the lowermost part of the thigh-bone of a man, or at least of some other animal' (Plot, 1676, p. 131). Plot believed that it was a real petrified bone, not a 'sport of nature', and that it had not been enlarged by petrification. It was thus clear that

> it must have belong'd to some greater animal than either an Ox or Horse; and if so (say almost all other authors in the like case) in probability it must have been the bone of some Elephant brought hither during the government of the Romans in Britain.

But this explanation did not suit Plot, because many objections could be made against it, namely that Roman authors did not mention elephants in Britain, that tusks had not been reported, and that large bones were often found in churches, an unlikely place to bury an elephant. For instance, after the Great Fire of London, in 1666, a large thigh-bone, 'supposed to be of a Woman', had been found when pulling down an old church during the building of a market-place; the bone was now to be seen at the King's Head tavern in Greenwich and Plot went on to remark:

> Now how Elephants should come to be buryed in Churches, is a question not easily answered, except we will run to so groundless a shift, as to say, that possibly the Elephants might be there buryed before Christianity florish'd in Britain, and that these Churches were afterward casually built over them.

Moreover, Plot had seen a living elephant in Oxford in 1676, and found its bones much bigger than the specimen from Cornwell. He thus came to the following conclusion:

> If then they are neither the bones of Horse, Oxen, nor Elephants, as I am strongly persuaded they are not, upon comparison, and from their likes found in Churches: It remains, that (notwithstanding their extravagant magnitude) they must have been the bones of Men or Women: Nor doth any thing hinder but they may have been so, provided it be clearly made out that there have been Men and Women of proportionable stature in all ages of the World, down even to our own days.

He went on to list Biblical, mythological and historical giants (including Theutobochus), and to mention the giants seen by travellers 'about the straights of Magellan' and near the Rio de La Plata.

Plot's giant thigh-bone has disappeared, but the illustration he gave of it enabled John Phillips to identify it as the distal end of the femur of a large megalosaur or a small cetiosaur (Phillips, 1871). It is now usually accepted that this first dinosaur bone to be described belonged to a Jurassic megalosaur (Halstead, 1970; Charig, 1979). It was later designated as *Scrotum humanum* by R. Brookes (1763) because of its shape, but this correct Linnaean name unfortunately must be considered a *nomen oblitum*, because it was not used by any other author (at least in this sense) for more than fifty years (Halstead, 1970).

After the end of the seventeenth century, belief in giants seems to have waned, at least in learned circles, and large fossil bones were less frequently attributed to them. Nevertheless, in 1754, the Spanish Franciscan monk Torrubia still devoted a part of his book on the natural history of Spain to a 'Spanish gigantology'.

While many bones of large extinct vertebrates seem to have been attributed to human giants, in some instances popular imagination preferred to identify fossil vertebrate remains as those of another class of mythical beings, the dragons. There are several well-documented cases of such interpretations, one of the best known being that of the Klagenfurt 'Lindwurm', which has been studied in detail by Abel (1925). In one of the squares of the city of Klagenfurt, in Carinthia (southern Austria), stands a monument which shows Hercules facing a quadrupedal winged monster bearing all the attributes of the classical dragon of western folklore. The dragon is the work of the sculptor Ulrich Vogelsang, who started working on it in 1590; the whole monument was not completed until 1636, 31 years after Vogelsang's death. The interesting point is that the head of the dragon is not entirely a product of Vogelsang's imagination: he used as a model the skull of a woolly rhinoceros (*Coelodonta antiquitatis*), which had probably been found about 1335 in a Pleistocene terrace near Klagenfurt, and was displayed for a long time in the town hall (it is now kept in the local museum). This rhinoceros skull was generally supposed to be that of the dragon of a local legend, and some of its features are recognisable on Vogelsang's sculpture. Othenio Abel has therefore dubbed the 'dragon monument' of Klagenfurt 'the oldest palaeontological reconstruction'.

Other dragon tales were associated with remains of cave bears (*Ursus spelaeus*), found in great abundance in some central European caves. Some seventeenth-century authors still interpreted such remains as those of dragons, in articles published in learned journals. The publications of the Kaiserliche Leopoldinische Akademie, an important German learned society of the time, for instance, contain reports on dragon bones from caves in the Carpathians and Transylvania by J. Paterson Hain (1672) and Heinrich Vollgnad (1673). Hain, a physician, gave a drawing of a 'dragon skull' which is easily recognisable as that of a cave bear. Among the many caves where dragon bones were supposedly to be found, the Drachenhöhle ('dragon cave') near Mixnitz, in Styria (Austria), is probably the most famous, because of the palaeontological excavations conducted there by Othenio Abel just after the First World War. A seventeenth-century map already mentions the cave as a place where dragon bones were found and collected (probably for medicinal purposes). At the time of the scientific excavations in 1923, stories about the Mixnitz dragon were still being told in some farms in the area; it was supposed to have been killed by a young local hero (Abel, 1939a).

Whether possible finds of Mesozoic reptiles, for instance in the bituminous shales of the Holzmaden area of Württemberg, where complete skeletons are relatively common, may have contributed to local legends about dragons, as suggested by Abel (1939a), is uncertain, as there are no definite records of such early finds.

Fossil bones were also associated with the legendary unicorn. The myth of the unicorn is probably of eastern origin and was initially based on reports about the Indian rhinoceros, but it soon became part of the European folklore. According to Chinese traditional medicine, the horn of the rhinoceros is reputed to have curative, and more particularly aphrodisiac, properties (this, incidentally, is one of the main reasons why rhinoceroses are fast becoming extinct in Africa as well as in Asia). The medicinal value of 'unicorn' horns became well known in Europe during the Middle Ages, and unicorn was often mentioned in pharmaceutical recipes until the eighteenth century. As true rhinoceros horns were exceedingly rare in Europe, a substitute was found for them in the form of fossil bones and teeth of extinct mammals, which became known as *unicornu verum* — whereas narwhal teeth were called *unicornu falsum* and considered as a mere surrogate, although the narwhal tusk soon became the typical attribute of the horse-like unicorn of western art. On account of its supposedly wonderful properties, *unicornu verum*, or 'fossil unicorn', was a very expensive kind of medicine, worth its weight in gold. It was considered especially effective against poison, bites and wounds, and was eagerly sought after. The main sources of 'fossil unicorn' were the caves in which remains of cave bears could sometimes be found in very large quantities. In some of these caves the resource was regularly exploited. In the Drachenhöhle at Mixnitz, mentioned above, excavations were started at least as early as the seventeenth century to collect the valuable *unicornu verum*. A cave in the Harz mountains of central Germany became known as the 'Einhornhöhle', or 'Unicorn cave' (Abel, 1939a).

Fossil bones from other kinds of deposits were sometimes also considered as 'fossil unicorn'. When abundant mammoth remains were found by a soldier in a loess deposit at Cannstatt, near Stuttgart, in 1700, Duke Eberhard Ludwig of Württemberg ordered that excavations should be conducted there (Wendt, 1971). These systematic excavations, which lasted for six months, may be considered as the first of their kind in the history of palaeontology, and they were the occasion for a renewed controversy concerning the nature and origin of the bones, although a chemical analysis performed by the physician J.S. Kerl showed that they were indeed organic remains (Wendt, 1971). More than sixty mammoth tusks were found, but only the best-preserved specimens were kept, the remainder being sent to the ducal pharmacy to be used as *unicornu verum*. Some of the fossils were even sent as a present of 'fossil unicorn' to the city of Zürich (Abel, 1939a). The myth of the unicorn was still vivid enough at the end of the seventeenth century and the beginning of the eighteenth to influence early reconstructions of fossil vertebrates, as will be seen in Chapter 2.

Although the old 'mythical' explanations of fossil bones were gradually replaced by more modern concepts during the eighteenth century, they persisted for a long time in the popular imagination. In 1873, for instance, an Abbé Canéto, of Auch, in south-western France, described a lower jaw of *Deinotherium* which had been found in 1838 in the bank of a small stream in the foothills of the Pyrenees. The fossil had been known to the local people for some time, and the tusks were thought to be *the devil's horns*. Finally, a young man who lived nearby broke them with a hammer to show that they were harmless — and Canéto could only collect whatever fragments were left (Canéto, 1873).

Other fossil vertebrates were thought to be holy rather than devilish. In 1843 a number of large bones were found when a field was being ploughed near Kishinev, in Bessarabia. The skeleton was roughly put together by a shepherd, and excited the enthusiasm of the local peasants, who sang and danced around the 'reconstruction', apparently thinking that the remains were those of a saint. The Russian governor of Bessarabia, however, thought otherwise: to him, the remains were nothing but those of an old Roman soldier with unusually large molars. Finally, the clergy decided that the whole affair was becoming something of a scandal, and ordered that the bones should be reburied. A few months later, the palaeontologist Alexander von Nordmann travelled to the place, but all he could find, with the help of an old woman, was a piece of rhinoceros jaw, now kept at the University of Helsinki (Nordmann, 1858–60; Fortelius and Kurten, 1979).

An interesting instance of a pseudo-historical interpretation of fossil bones, somewhat reminiscent of Mazurier's Theutobochus, was reported by J. Miquel in 1896. A rich late Eocene vertebrate locality near the village of La Livinière, in the *département* of Hérault (southern France) was locally known as 'Alaric's grave', after the famous Visigoth king who occupied the area in the fifth century AD, and the abundant mammal and reptile bones which were encountered there were thought to be the remains of defeated Visigoth warriors.

Outside Europe, other civilisations also knew about fossil vertebrate remains and had explanations of their own for these unusual objects. Well-preserved fossil fishes were correctly interpreted as petrified animals by Chinese authors as early as the twelfth century AD (Edwards, 1976); at the time, the demand for such fossils was so considerable that the local people had already started to make forgeries by painting the outlines of fishes on stones. However, the most famous Chinese vertebrate fossils were undoubtedly the so-called 'dragon teeth' (*long chi*) and 'dragon bones' (*long gu*), actually the remains of Neogene and Pleistocene mammals, which were supposed to possess medicinal properties, and are still to be found in some traditional Chinese drugstores, as the author ascertained in Bangkok a few years ago. In contrast to its Western counterpart, the Chinese dragon was thought to be a benevolent creature, and was said to make rain fall from the clouds. Bones and teeth of dragons were supposedly found when these beings had been unable to fly back to heaven because of lack of clouds. According to an eighteenth-century medical treatise, dragon bones could be used to cure diseases of the heart, kidneys, intestines and liver. They were prescribed against a variety of illnesses, from constipation to nightmares and epilepsy. They could be taken raw, fried in fat, or cooked in rice wine. Dragon horns (antlers of fossil deer) and dragon teeth were used in the same way, and were also supposed to be efficient against heart diseases and epilepsy (Abel, 1914). There was of course much demand for such wonderful medicine, and large quantities of fossil bones and teeth were dug out of Pleistocene caves or Neogene outcrops and put on the market: a report of the Imperial Chinese Customs Administration for 1885 indicates that 20 tonnes of dragon bones were loaded in Chinese ports in a single year. Today, and despite some control exercised by palaeontologists, several tonnes are still exported every year (personal communication of professor Li Chuan-kuei, Beijing). The price of dragon bones and teeth varied according to their quality; teeth were more expensive than bones. As will be seen in Chapter 9, the scientific interest of this peculiar kind of medicine was recognised by Europeans in the nineteenth century, and this led to important palaeontological discoveries in the drugstores of large Chinese cities. How the dragon bones and teeth were obtained, however, and where exactly they came from, remained something of a mystery to Western palaeontologists until the twentieth century, when it finally became possible to visit some of the fossil localities. Some of the dragon teeth were those of representatives of the so-called *Hipparion* fauna, of Neogene age. Others came from Pleistocene fissures or cave deposits. In the 1920s the American palaeontologist Walter Granger had the opportunity to spend several winters at such a 'dragon bone' site in Sichuan Province (Granger, 1938). Near the small village of Yenchingkou, above the gorges of the Yangzi River, vertebrate remains occurred in sink-holes on a ridge of Palaeozoic limestone. The sink-holes had acted as natural traps in which Pleistocene animals had become entombed. The holes, which could be 50 feet (15m) deep or more, were exploited by local farmers living on the ridge, who tilled the soil in summer, and excavated the fossiliferous pits in winter after the harvest. The bones were taken out of the pits by men lowered into them with ropes rigged to a pulley. The sediment at the bottom of the pits was raised to the surface in wicker baskets, and searched for bones. The bones were cleaned and stored in farmhouses where they were left to dry until wholesale merchants came to buy them.

The Chinese had also heard about the abundant mammoth remains found in the frozen ground of Siberia, and they used mammoth ivory for carving. Influenced by the legends of the Siberian tribes, they believed that the mammoth was a kind of gigantic mole which lived underground. According to a sixteenth-century natural history book quoted by Cuvier (1836, p. 124), the animal was named *tien-schu*, or *tun-schu*, that is, 'the mouse which hides':

> It stays continually in underground caves; it looks like a mouse, but its size equals that of an ox or a buffalo. It has no tail; its colour is dark. It is very strong, and digs holes in rocky and wooded places.

The *tun-schu* was supposed to die as soon as it was exposed to the rays of the sun or moon. Interestingly, the idea that large bones found in the ground were those of burrowing creatures which actually lived underground also arose independently in South America: when Darwin, during the voyage of the *Beagle*, found bones of huge extinct mammals in the Pleistocene deposits of the Pampas, he was once told by local inhabitants that these remains were those of animals which lived in underground burrows, in the same way as the vizcacha, a common South American burrowing rodent.

2

Diluvialist Interpretations of the Seventeenth and Eighteenth Centuries

All the myths and wild speculations about giants and dragons mentioned in Chapter 1 contributed but little to a solution to the enduring problem of the nature of fossils. The question was made even more complex by the fact that under the term 'fossil' were included widely different objects: not only organic remains, as in the present meaning of the word, but also minerals, strangely shaped stones and concretions; a fossil simply was any object dug up from the ground, and it was not until the late eighteenth century that the use of the term became generally restricted to remains of organisms or direct traces of their activity. Before that, few authors were really concerned about separating 'organic' from 'inorganic' fossils.

Ever since the Greeks, the main question had been whether those fossils which looked like organisms were the petrified remains of ancient living beings, or the result of mysterious processes taking place inside the earth. What the 'plastic virtue' or *vis plastica* responsible for the formation of fossils could be was by no means clear; nevertheless, this rather unsatisfactory explanation was accepted, under various guises, by a number of scholars of the Middle Ages and the Renaissance, and still found supporters at the end of the seventeenth century and the beginning of the eighteenth. Large fossil bones, as shown in Chapter 1, often received a different explanation, as they were thought to have belonged to living, albeit unusual, creatures such as giants, dragons and unicorns. Nevertheless, some authors thought that even these bones could be formed inside the earth by the *vis plastica*. As already mentioned, in the sixteenth century the anatomist Fallopio (1523–62) believed that elephant remains found in Italy were mere earthy concretions (Edwards,

1976). Even Riolan, who did not believe in giants, could not make up his mind as to whether the bones of the so-called Theutobochus were those of an elephant or 'sports of nature'. The idea of a plastic force at work in nature to produce shapes similar to living organisms, or parts thereof, still found supporters in the eighteenth century. In 1768 the Frenchman J.B. Robinet produced a curious theory, according to which nature 'learned how to make man' by producing fossils shaped like various organs of the human body. He believed that fossils lived a kind of embryonic life, 'well below that of sleeping animals and of plants'. To illustrate his assertions, he illustrated and described many stones and fossils more or less reminiscent of human organs. One of them was Plot's fragmentary dinosaur femur, which Robinet, taking Brookes's denomination literally, described as a 'stony scrotum' showing many anatomical details of real human testicles (Buffetaut, 1980). By the second half of the eighteenth century, however, such bizarre conceptions were becoming minority views.

Since the Middle Ages the opinion according to which fossils were the remains of once living beings, had been slowly gaining ground. Among the most famous proponents of such a concept were Albertus Magnus (1193?–1280), Leonardo da Vinci (1452–1519), Girolamo Fracastoro (1483–1553) and Bernard Palissy (1510–90) (for details about their ideas, see Furon, 1951; Rudwick, 1972; Edwards, 1976; Faul and Faul, 1983). These authors, however, were mainly concerned about fossil invertebrates, more abundant than vertebrates and leading to far-reaching conclusions about the past distribution of land and sea. From time to time, nevertheless, the resemblances between especially well-preserved fossil vertebrates and living animals did attract some attention. During the seventh crusade of 1248, for instance, King Louis IX of France was shown some fossil fishes from the famous Cretaceous localities of Lebanon. In the words of Joinville, who wrote a chronicle of the king's life (Furon, 1951),

> When the king was in Sayette (Sidon), a stone was brought to him, which split into scales, the most wonderful in the world, for, when a scale was lifted, the shape of a fish was found between the two stones. The fish was stony, but it lacked nothing of its shape, neither eyes, nor bones, nor colour, nor

anything else that it might have had, had it been alive. The king gave me a stone, and I found a tench inside, of brown colour, and such as a tench should be.

Fish teeth were actually among the first vertebrate fossils to receive a more or less correct interpretation, when it was realised that the glossopetrae were nothing but shark teeth. In 1558 the famous naturalist Conrad Gesner (1516–65) had already commented on the resemblance between glossopetrae and the teeth of sharks, and had published an illustration to make his point (Rudwick, 1972). Final demonstration came in the seventeenth century, when several authors working in Italy all came to the same conclusion (their works on the subject have recently been reprinted by Morello, 1979). The first of them seems to have been Fabio Colonna (1567–1650), who in 1616 published a treatise on glossopetrae in which he showed that they were actually shark teeth, and were often found together with shells of marine molluscs (which he also took to be remains of living beings).

More than fifty years later, Steno was again to demonstrate that the 'tongue-stones' were the fossilised teeth of sharks. Steno, born Niels Stensen (1638–86), a Dane who had studied anatomy in Copenhagen, and later in the Netherlands and in France, became famous for his careful anatomical dissections, which attracted a large public (Hoch, 1985). In 1665 he moved to Florence, where Duke Ferdinand II gave him a position in a hospital, which gave him the opportunity to pursue his scientific researches. Soon after his arrival in Tuscany, a large shark was brought ashore by fishermen near the town of Livorno, and its head was taken to Florence by order of the duke, to be dissected there by Steno. The result was an anatomical study published in 1667, which included, besides a description of the myology of the shark's head, a discussion of glossopetrae. Examination of the teeth had made it obvious to Steno that the so-called 'tongue-stones' could be nothing else but the petrified teeth of sharks. Whether Steno knew of Colonna's work on the same topic is uncertain — in any case, he did not mention it (Rudwick, 1972). His demonstration included a discussion of the formation of glossopetrae in which he showed that they did not grow within rocks, but gave all signs of having been buried in soft 'earth' mixed with water. The complete similarity of shape between glossopetrae and the teeth of a living shark was final proof of their basic identity, and an upheaval of some kind could explain why fossil shark teeth could be found well above sea-level. These more geological questions also were of interest to Steno, who has been acclaimed as a precursor of stratigraphy because of his famous *De solido intra solidum naturaliter contento dissertationis prodromus*, published in Florence in 1669, in which he clearly expressed the 'law of superposition', according to which a rocky layer is always younger than the one it overlies. Reconciling his geological views with a literal interpretation of Genesis was a difficult task for Steno, who had converted to Catholicism after arriving in Tuscany. In 1675 or 1676 he became a priest and abandoned his scientific activity, so that the complete 'dissertation' of which the *Prodromus* was supposed to be a forerunner was never written.

A few years after Steno's description of the shark's head, a Sicilian painter and naturalist, Agostino Scilla (1639–1700) published a book on fossils in which he attempted to use common sense to fight vain speculations. Scilla (1670) was convinced that fossils were the remains of living beings and not the result of a mysterious growth within the earth. Among the Tertiary fossils from southern Italy and Malta which he used in his demonstration were glossopetrae, which he correctly identified by comparison with the teeth of living sharks.

Steno's work apparently exerted more influence than either Colonna's or Scilla's, possibly because Steno had more contacts with naturalists outside Italy. It is known (Hoch, 1985) that, when in Montpellier, he had met some English scientists; among them John Ray (1627–1705) and Martin Lister (1639–1712), a physician from York who had described many fossil shells but did not believe in their organic origin. There was certainly some exchange of ideas between Steno and the newly founded Royal Society. The meaning of fossils was one of the important topics discussed by the members of the Society (Rudwick, 1972; Faul and Faul, 1983), but, not unexpectedly, the controversy centred mainly around fossil invertebrates, which were more abundant and more easily recognised than vertebrate remains. Robert Hooke (1635–1703) drew a sharp distinction between 'petrified' remains of animals or plants and crystals, and believed that many species of former ages had become extinct, while others had appeared (Faul and Faul, 1983). He strongly opposed the 'plastic virtue' hypothesis which was still accepted by many of his colleagues, including Martin Lister, mentioned above. John

Ray, although impressed by evidence for the organic origin of fossils, was troubled by the lack of identity between many of them and living forms, which strongly suggested that some species had become extinct. The suggestion that whole species had been allowed to disappear implied that Creation had not been perfect, and this seemed to contradict the Christian doctrine. There were two possible ways out of the dilemma: either the apparently 'extinct' forms were still alive somewhere on earth, although undiscovered, or fossils were not, after all, the remains of once living organisms, but mere inorganic formations. Ray finally chose to 'sit on the fence', and concluded that some fossils (including ammonites) were inorganic, whereas others were petrified remains of animals and plants (Rudwick, 1972).

Thus, at the end of the seventeenth century, the problem of extinction, which was to prove a major obstacle to the development of vertebrate palaeontology (and of palaeontology as a whole) was already at the centre of the controversy about the meaning of fossils. Ray's friend Edward Lhwyd (1660–1709), devised a peculiar explanation according to which fossils grew *in situ* in the rocks from the 'seeds' of living organisms which had been washed into the earth through cracks in the rocks. Among the many fossils described and illustrated by Lhwyd were ichthyosaur vertebrae, which he compared with fish vertebrae (Howe, Sharp and Torrens, 1981).

The general controversy about the organic or inorganic nature of fossils thus also extended to fossil tetrapods. There was a relative progress of comparative anatomy during the second half of the seventeenth century, with dissections of various animals being performed and described by such scientists as Claude Perrault (1613–88) in France and Tyson (1649–1708) in England. This led to a better knowledge of the osteology of many vertebrates, which in turn made it easier to show that the fossil bones which kept turning up in various parts of Europe could not have belonged to giant men. Thus, when in 1688 some enormous bones were found at Vitorchiano in Italy, the anatomist Campini decided to compare them with elephant bones. As no elephant skeleton was available in Rome, he had casts made from a specimen in the Medici collection in Florence. He found a striking resemblance between the bones from Vitorchiano and those of the elephant; this led him to the conclusion that all the so-called giant bones in Italian collections had in fact belonged to elephants (Langenmantel, 1688; Archiac, 1864).

Such ideas about the real affinities of fossil skeletons gained ground slowly. When in 1696 a mammoth skeleton was found at Gräfentonna, near Gotha in Thuringia, the medical college of Gotha decided that it was a mere sport of nature, a mineralogical object formed in the earth. However, Wilhelm Ernst Tentzel, historiographer and librarian of the Duke of Saxony, followed Campini and demonstrated that the bones were those of an elephant in a dissertation which was sent to the Royal Society, together with some of the remains (Tentzel, 1697). Although a controversy developed between Tentzel and the Gotha physicians, the idea that the large fossil bones often encountered in the ground could be those of animals was becoming acceptable to a growing number of scholars. What these animals had been was not yet clear, as shown by the interpretations given to another German find. In 1663 workmen digging in a limestone quarry on the Zeunickerberg at Quedlinburg, came across the skeleton of an enormous animal in what was apparently a karstic pocket. The bones, which belonged to a mammoth, and probably a rhinoceros (Abel, 1939a), were taken to a local monastery, and they soon attracted the attention of various scholars. Among them was Otto von Guericke, the famous mayor of Magdeburg and inventor of the vacuum pump, who mentioned the find in one of his works and prepared a reconstruction of the skeleton, which he sent to the philosopher G.W. Leibniz (1646–1716). The drawing was reproduced in Leibniz's *Protogaea*, a posthumous work which was published by C.L. Scheidt 33 years after the philosopher's death. It clearly shows how much von Guericke was still influenced by the old ideas about 'fossil unicorn': the somewhat horse-like skull displays a long straight horn jutting out of its forehead. The rest of the body, however, is more unusual, as there are no hind limbs. According to Leibniz, the Quedlinburg skeleton possessed a horn 'five ells long' and the width of a man's thigh at the base. Leibniz thought that this animal was probably a terrestrial beast, but he interpreted many other remains of fossil vertebrates as those of marine animals such as walruses. He had no doubt about the organic nature of fossil fishes: the 'petrified' fishes of the Permian *Kupferschiefer* of Germany were the remains of true fishes, and the glossopetrae, whether from Malta or from Lüneburg (in northern Germany), were the teeth of sharks. However, Leibniz still echoed the traditional tales about the healing virtues of fossils when he remarked that, although their properties were much exaggerated, the glossopetrae could be useful as dentifrice, 'either because the powder made from them is sufficiently hard and abrasive, or because this dental substance seems to be the less harmful for the teeth'. Leibniz clearly recognised that some fossils (such as ammonites) had no known present-day counter-parts, and he put forward some explanations for such a remarkable fact: such fossils may have been transported from remote, still unexplored places (and marine life was still far from being completely known). After all, the New World was yielding many hitherto unknown living beings. Moreover, it was conceivable that 'in the great changes the world has gone through, many animal forms had been transformed'. According to Leibniz, the most important of these changes had been the Biblical Flood. When he mentioned the bones found in the Baumann and Scharzfeld caves of the Harz mountains, he inter-preted them as 'undamaged bones of sea-monsters and other animals from an unknown world, which cannot have been born in these places, but have been transported inside the ocean by the violence of the waters'. The waters in question were those of the Flood, which had supposedly washed the animal remains into the caves.

Leibniz was neither the first, nor the only author to think that the Flood had been responsible for the formation of fossil localities. This idea had already been put forward in the Middle Ages by Ristoro d'Arezzo, an Italian monk who in 1282 had claimed that fossil shells found in mountains had been brought there by the Flood. This hypothesis had been challenged as early as the fifteenth century by Leonardo da Vinci, but this had not prevented it from becoming popular. The diluvialist theory was attractive enough, because it reconciled the idea of the organic nature of fossils with Scripture: if petrified remains of marine and other organisms were found far inland, sometimes on high mountains, it was because they had been carried there by the waters of the Flood, as mentioned in the Bible — and there was no longer any need to invoke a mysterious *vis plastica* to account for the formation of fossils within rocks. Although many later authors have passed harsh judgements on diluvialist theories, considering them as disastrously misleading concepts which severely hampered the progress of palaeontology, it should be admitted that such ideas at least had the merit of making the organic origin of fossils more easily acceptable to those who did not question the literal truth of the Bible. The Flood was a major event of earth history, and its reality could not be doubted. If fossils could be accepted as one of its results, their origin no longer posed any major problem — at least as long as it was not recognised that some of them had no living counterparts.

Interest in the Flood and its geological and biological consequences had much increased in the last decades of the seventeenth century, when several British scholars, most of them members of the Royal Society, started to publish 'theories' or 'systems' of the earth. Supposedly, they were reconstructions of the early history of the world, but their basis of scientifically ascertained facts was extremely flimsy, and they consisted mainly of speculations drawn from the story of Creation as told in Genesis (Rudwick, 1972; Faul and Faul, 1983). All these theories, of course, stressed the importance of the Flood as a major event in the history of the earth. Thomas Burnet's *Sacred Theory of the Earth*, first published in Latin in 1681, was one of the most famous and debated of these speculative works. According to Burnet, the surface of the earth had once been smooth, without either mountains or oceans. The Deluge had been the result of fracturation and collapse of the outer crust of the earth, which had liberated waters previously hidden beneath this crust. In this interpretation, mountains were the mere ruins of the broken crust, and as pointed out by Rudwick (1972), it was not possible to consider marine fossils found in mountains as the remains of sea animals as there had been no oceans on the primeval earth.

John Woodward's *Essay Toward a Natural History of the Earth* (1695) was more progressive, from a palaeontological point of view, in that it was in agreement with an organic origin of most fossils. According to Woodward (1665–1728), at the time of the Deluge all solid materials, including organic remains, had been brought into some kind of suspension in the waters. Afterward, these materials had settled out in the order of their specific gravity, the heaviest ones at the bottom. This had given rise to the stratification observed in rocks. The organic remains had thus been enclosed in rock strata where they could now be found as fossils.

Woodward's *Essay* was to have a determining influence on one of the most famous of the early palaeontologists, the Swiss Johann Jakob Scheuchzer (1672–1733). Scheuchzer had studied medicine at the Altdorf Academy in northern Bavaria, and received a doctorate at the University of Utrecht, after which he had returned in 1694 to his native town of Zürich, where he became town physician. He was keenly interested in natural history, and his excursions in the Alps, then usually regarded as a horrible wasteland, resulted in valuable contributions to the geographical knowledge of Switzerland. Scheuchzer also maintained a large correspondence with many famous European scholars, including Leibniz and Hans Sloane (Fischer, 1973). He had started to collect fossils in 1690, but at first he was not convinced of their organic origin. After reading Woodward's *Essay*, however, he changed his mind. He was enthusiastic about this work, which he translated into Latin, and reinterpreted his vast fossil collection in 'diluvialist' terms. The reason for his enthusiasm was partly religious: if fossils were indeed remains of living beings buried in sediment at the time of the Flood, they could be used as material witnesses to the veracity of Scripture. One of Scheuchzer's first efforts in this direction was his famous *Piscium querelae et vindiciae* (1708), a slim but well-illustrated volume in which were presented the 'complaints and claims' of fossil fishes. Not only had these fishes fallen victim to the Deluge, for which they were not responsible, but their organic nature was disputed by an 'unsound philosophy' which explained all fossils as purely mineralogical formations. The sins of men had been the ultimate cause of their deaths when they had been left stranded by the ebb of the waters, and now men were denying them an organic origin. Despite the rather unusual form of this discourse, Scheuchzer's work reveals his knowledge of a fairly large number of fossil fish localities: he mentions not only the famous Miocene fishes from Oeningen, on the German side of Lake Constance, but also those from the Oligocene Glarus shales of Switzerland, and those from the Kupferschiefer of Germany. The Maltese glossopetrae are interpreted as the teeth of sharks which died during the Flood. Fossil fish teeth are also reported from many other places, including Lüneburg, Sheppey and Syria. Scheuchzer even mentions teeth from Maryland and the Carolinas, and this seems to be one of the earliest reports of fossil fishes from North America. The book also contains a mention of a supposed 'crocodile' from the Kupferschiefer (in fact a Permian reptile). More interesting still to a diluvialist such as Scheuchzer, were 'two petrified human dorsal vertebrae from Altdorf'. These had been found near the gibbet of the town, but Scheuchzer was convinced that they were not those of a recent victim. For him, they were the remains of an antediluvian sinner who had been drowned in the Flood. On this point, Scheuchzer was in complete

27

disagreement with one of his former comrades at the Altdorf Academy, Johann Jakob Baier, who in his *Oryctographia Norica* (a description of the fossils of the Nuremberg region, also published in 1708) had identified the vertebrae as those of a fish (they in fact belonged to a Liassic ichthyosaur). In a letter to Scheuchzer (Hölder, 1960), Baier, who also believed in the organic nature of fossils, tried to explain to him that comparison with human vertebrae revealed a number of differences, which made it quite impossible to identify the Altdorf vertebrae as those of a man. Scheuchzer, however, was unmoved. As he remarked in 1731 in his *Physica sacra* (a sort of 'scientific' exegesis of the Bible, with remarks on Noah's Ark and figures of fossils considered as victims of the Flood), what really bothered him was the remarkably small number of human fossils hitherto found. After all, men had been the cause of the Deluge, which had claimed so many innocent victims among animals, and it was strange that up to 1725 the only known fossil human remains should have been the two vertebrae from Altdorf. In 1725, however, the situation changed with the discovery of the skeleton of one of those 'miserable sinners' in the Oeningen limestone (Jahn, 1969). In a letter to Hans Sloane published in the *Philosophical Transactions of the Royal Society* in 1727, Scheuchzer grew enthusiastic about this new discovery (translation from the German version in Hölder, 1960, p. 365):

> We have obtained some remains of the human race which was destroyed by the Flood. Until recently, I possessed in my fairly large collection only two blackish shining petrified dorsal vertebrae of such a man. Now, however, my museum has been enriched by the happy acquisition of remains embedded in the Oeningen shale which seem to be worthy of consideration. What can be recognised there is not the product of imagination, but numerous really preserved parts of a human head, easily distinguishable from all other animal genera . . .

This *Homo diluvii testis*, 'man, witness of the Flood', was, according to Scheuchzer, an incontrovertible proof of the reality of this part of the Biblical record, older and more demonstrative than all the monuments of Antiquity. As shown by Cuvier about a century later, this 'witness of the Flood' was in fact the partial skeleton of a giant salamander. To Scheuchzer, however, it was both a remarkably demonstrative proof of the reality of the Flood and a moral lesson to present-day sinners.

Even though Scheuchzer's 'fossil man' did not become generally accepted (in 1783, for instance, Jean-Etienne Guettard reviewed supposed human fossils, and concluded that they were not truly fossil men), his interpretation of fossils in general as remains of victims of the Biblical Flood was followed by many eighteenth-century authors, and still found echoes well into the nineteenth century. The careful collection and description of fossils was no longer an idle pastime; it became a legitimate way to confirm the veractiy of Scripture by direct evidence found in nature.

Some eighteenth-century authors, however, were not as much obsessed by the Flood as Scheuchzer, and they gave more sober descriptions of the fossils which were now accumulating in large numbers in some collections. A good example is provided by a remarkable work published by two Bavarians, Georg Wolfgang Knorr and Johann Emmanuel Walch, under the title *Sammlung der Merkwürdigkeiten der Natur, und Altertümer des Erdbodens, welche petrificierte Körper enthält* (Collection of wonders of nature and antiquities of the earth, comprising petrified bodies). This four-volume book, printed in Nuremberg between 1755 and 1778, contained magnificent coloured plates prepared by Knorr himself, who was both an artist and a naturalist. The text was mainly the work of Walch, who completed the book after Knorr's death. Their book is mostly descriptive, and few 'osteoliths', as vertebrate fossils were then called, were illustrated, because, according to Walch, they were considered as less interesting than other 'petrifications'. Nevertheless, glossopetrae from Malta, England and Germany are shown and described, as well as complete fishes from the Permian of Eisleben, the Jurassic of Solnhofen and Pappenheim, and the Miocene of Oeningen. There is comparatively little about fossil tetrapods: an elephant tooth from Basle, the horn of a large ox, and a fragment of a mastodon tooth. Bones and teeth from the famous Baumann's cave in the Harz, already mentioned by Leibniz, are described; these remains of cave bears were attributed with some hesitation to the cetacean *Orca*, the killer-whale — which, of course, was reminiscent of Leibniz's interpretation of these fossils as bones of 'sea-monsters' washed into the cave by some violent inundation (in all likelihood the Flood). Although the book by Knorr and Walch contains relatively little speculation, there are interesting remarks about the duration of the 'catastrophes' which have affected the earth's surface; they are supposed to have lasted several thousand years, although too-high estimates are rejected. Interestingly, Walch realised that all fossils were not contemporaneous; this implied that they could not have been formed by a single event.

One inescapable result of the growth of fossil collections and of the publication of descriptive catalogues was that, in many instances, it proved difficult or impossible to identify fossils, because no living counterparts could be found. The debate about extinction was to dominate eighteenth-century palaeontology.

3

Eighteenth-century Philosophers and the Problem of Extinction

With the development of anatomical knowledge, it slowly became possible not only to reject the old stories of dragons and giants, but also to attempt to identify the animals whose fossil remains were found in the rocks. This quickly resulted in theoretical difficulties, as many of these creatures seemed to belong to exotic species no longer found in the temperate climates of Europe, or even to completely unknown forms for which no living counterparts were known. In the case of marine organisms, such as ammonities, it could still easily be supposed that extant representatives were living in the depths of the sea, where it would be difficult to capture them. As far as land vertebrates were concerned, however, the problem was more difficult to solve. Of course, large parts of North and South America, Africa, Asia and Australia were still unexplored, and many unknown species could (and did) live in these regions. Nevertheless, by the end of the eighteenth century many naturalists were beginning to feel confident that most living species of large terrestrial vertebrates were already known to science, and to suspect that some fossil bones may have belonged to no longer extant species. This, of course, was not easy to admit, as it implied that God had allowed some parts of His supposedly perfect Creation to perish. The Biblical Flood did provide an explanation for the deaths of uncounted numbers of *individual* animals, but it was not supposed to have caused the complete extinction of any species. After all, Noah's Ark had been there to save representatives of all animal species. Local extinctions were easier to envisage, although their causes were problematical, but worldwide extinction of an entire species was a concept which most eighteenth-century scientists found difficult to accept. It

31

was not until the very end of the century that the reality of extinction finally was demonstrated.

As early as 1697, Thomas Molyneux, an Irish scholar who founded the Dublin Philosophical Society, had tackled the problem of extinction in his 'Discourse concerning the large horns frequently found under ground in Ireland . . .' These were the antlers of the famous 'Irish Elk'. Three skulls had been found in a white marl below a peat-bog by a Mr Henry Osborn from Dardistown, near Drogheda, in County Meath. Molyneux came to the conclusion that these skulls showed 'that the great American Deer, call'd a Moose, was formerly common in . . . Ireland'. Reports on the moose had been sent from New England by British colonists, and Molyneux had no doubt that the Irish fossils could be ascribed to the same species. By identifying the Irish Elk with the moose, Molyneux had no need to assume that the Irish species had become completely extinct. He discussed the question of extinction at some length (Molyneux, 1697, p. 489):

> That no real Species of Living Creature is so utterly extinct, as to be lost entirely out of the World, since it was first created, is the Opinion of many Naturalists; and 't is grounded on so good a Principle of Providence taking Care in general of all its Animal Productions, that it deserves our Assent. However great Vicissitudes may be observed to attend the Works of Nature, as well as Humane Affairs; so that some entire Species of Animals, which have been formerly Common, nay even numerous in certain Countries, have, in Process of Time, been so perfectly lost, as to become there utterly unknown; tho' at the same time it cannot be denied, but the kind has been carefully preserved in some other part of the World.

It remained to be explained how the moose had become extinct in Ireland, and Molyneux, who did not believe its remains were antediluvian, had a ready and sensible answer: it had been exterminated by man. This 'overkill' hypothesis was to be applied to the extinction of other large fossil mammals by other authors during the eighteenth century.

Among the unexpected vertebrate remains which were unearthed in eighteenth-century Europe were those of reptiles reminiscent of tropical forms. In 1710 Christian-Maximilian Spener, a Berlin physician, described a reptile skeleton which had been found in a copper-mine near Eisenach at a depth of nearly 100 feet (30m). Spener suggested that the animal may have been a crocodile. A few years later in 1718, a second specimen from the same locality was described by H. Linck, a pharmacist from Leipzig, who also thought it was a fossil crocodile. A third fossil reptile from the Kupferschiefer of Germany received a stranger interpretation when it was described by Emmanuel Swedenborg in 1734. In a treatise on copper, the famous Swedish mystic, who was also interested in geology, identified the specimen as a *Meer-katze*, a 'sea-cat' — which is the German name for a monkey. What Swedenborg really meant by this is uncertain; most authors admit that he meant a monkey, but it has also been suggested that he thought that the specimen was a seal (see Cuvier, 1836, t. 10, p. 105). According to Wendt (1971), what he had in mind was not a real animal but some kind of supernatural being. These Kupferschiefer reptiles, after being interpreted by Cuvier as monitor-like reptiles, have finally been identified as primitive Permian reptiles.

In 1758 a true crocodilian was discovered in Liassic rocks (the so-called 'Alum Rock') on the Yorkshire coast near Whitby. Independent reports on this fossil were published in the *Philosophical Transactions*. In the first one, William Chapman gave a brief description of the skeleton, and concluded that it was an animal 'of the lizard kind', and that because of its length (more than ten feet), it seemed to have been 'an allegator'. Somewhat later, a more careful description was given by a Mr Wooller, who noted that the fossil had been found together with ammonites. Two years earlier, in 1756, George Edwards, a London physician, had described a peculiar new kind of crocodile in the *Philosophical Transactions*. The specimen, newly hatched, had been sent to him from Bengal, and one of its most salient features was its long, beak-like snout (this obviously was a young gavial from the Ganges). Wooller considered that the Whitby fossil (a teleosaurid crocodilian) resembled this Indian crocodile 'in every respect'. He also remarked that the conditions of the locality and the disposition of the strata seemed (Wooller, 1758):

> clearly to establish the opinion, and almost to a demonstration, that the animal itself must have been antediluvian, and

that it could not have been buried or brought there any otherwise than by the force of the waters of the universal deluge.

This is fairly typical of the way in which some eighteenth-century naturalists strived to explain the occurrence of fossils of exotic animals in the rocks of European countries. The Flood, having been an extremely violent and worldwide event, accounted for the transportation and burial of carcasses even quite far from their places of origin. There was no real need to assume that tropical creatures had once lived in Europe; their remains had simply been brought there by the Deluge.

Tropical reptiles were not the only unexpected animals to be discovered in the rocks of temperate countries during the eighteenth century. Some finds seemed to indicate that inhabitants of the polar regions had once lived there as well. One of these finds in France received some publicity after the naturalist Jean-Etienne Guettard became interested in it. Guettard (1715–86) was one of the most remarkable French naturalists of the eighteenth century. He published extensively on various subjects, ranging from botany to zoology, but he is mainly remembered for his pioneering work in geology. He was the co-author with the chemist Lavoisier of an early kind of geological map (more accurately, a map showing the distribution of various rocks and minerals), and he discovered the extinct volcanoes of central France. He was much interested in fossils too, and wrote several monographs on them. One of these was devoted to bones found near Etampes, Guettard's native town. The story of the discovery, as told by Guettard in 1768, is as follows: about twenty years before, a workman quarrying sandstone blocks near Etampes (some 50 km (31 miles) south of Paris) found a number of bones in a fissure deposit. He thought he was digging in a graveyard, an idea which he found disquieting. He therefore decided to bury the bones under sandstone blocks. Fortunately, two remarkably large specimens attracted the attention of some local inhabitants, who collected one of them. The other one, which had been left in the quarry, was broken by children. A few months later during a visit to Etampes, Guettard heard about the find. He managed to obtain the remaining bone, and recognised that it was a fossil. He then visited the quarry, where more fossil bones were dug up in his presence. This convinced him of the great interest of this locality, and he reported it to his patron, the Duke of Orleans (Guettard was the curator of the Duke's natural history cabinet). The Duke was interested and gave his approval to Guettard's project of digging for the bones the quarryman had buried. However, the latter argued that the job was too dangerous and refused to do it. He could only be persuaded to put aside any bones he might find in the course of further quarrying, and then turn them over to collectors. Several bones, antlers and jaws were thus obtained. In 1751 Guettard showed some of them to the Royal Academy of Sciences, which entrusted him with the task of studying this material and trying to identify it. In 1754 however, a Mr Clozier from Etampes published a letter on the locality (reprinted in Alléon Dulac, 1763), in which he claimed that the carcass of a reindeer had been found there together with hippopotamus bones; the bones had purportedly been identified as such by the Academy of Sciences. Clozier speculated briefly about what could have brought together a reindeer from Lapland and a hippopotamus. Although he did not mention the Flood, he obviously thought that the bones had been carried where they had been found by the agency of water. Guettard was much surprised by this letter, mainly because the Academy had refrained from giving a precise identification, although resemblances with hippopotamus bones and reindeer antlers had been mentioned. Moreover, there was no evidence of a complete reindeer carcass; what had been found were the dissociated bones of several individuals.

Guettard was not the only one to take exception to Clozier's letter. No less a figure than Voltaire became interested in the bones from Etampes. In a small work on the 'singularities of nature', the famous writer expressed his doubts about the origin of fossils. He joked about marine shells found on mountains having been brought there by pilgrims on their way to Rome or Santiago de Compostela, and defended the idea that fossil shells grew in the ground. The reindeer and hippopotamus from Etampes also excited his sarcastic wit (Fallot, 1911, p. 217):

A few years ago, the bones of a reindeer and a hippopotamus were discovered, or thought to have been discovered, near Etampes, whence it was concluded that the Nile and Lapland had once been on the road from Paris to Orleans. One should rather have suspected that some curious person had once had in his collection the skeletons of a reindeer and a hippopotamus.

Voltaire's scepticism was not to the liking of many naturalists, however, and he was sharply rebuked by Buffon.

Guettard himself was sceptical about the occurrence of reindeer and hippopotamus remains near his home town. After comparing them with the bones of a doe, he concluded that most of the bones may have belonged to deer, and that there was no real need to suppose that reindeer had once lived in France. As to the large bones which had been attributed to a hippopotamus, Guettard remarked that such fossils had often been interpreted as the remains of exotic animals, especially elephants, brought to Europe by the Romans. He thought this explanation unlikely, and also rejected the idea that a change in the earth's position could have taken place, before which animals from tropical regions could live in temperate zones. He found it hard to accept that some species had completely disappeared, but he did admit that some could have become locally extinct, a position reminiscent of Molyneux's views on the Irish Elk. The large bones from Etampes may have belonged to a huge kind of wild ox similar to the *urus*, which Guettard thought was still alive in Poland (in fact, the last urus, or aurochs, not to be confused with the European bison, had been dead for more than a century at the time Guettard was writing). Such an identification made the hypothesis of extinct species unnecessary. The explanation which Guettard favoured was that the accumulation of animal bones found near Etampes was the result of pagan sacrifices: the ancient inhabitants of the Etampes region had sacrificed part of the game they hunted to their gods by throwing carcasses into caves or holes in the ground. Guettard had read many reports on large bones found in the earth, but he was extremely cautious about their significance. The main reason for his attitude was that he realised perfectly that 'comparative anatomy is not yet advanced enough, especially as far as skeletons are concerned, to bring to this matter all the light and clarity it requires, and may eventually receive' (Guettard, 1768, p. V). He did describe a number of fossil bones, several of them from the gypsum quarries near Paris, but he was uncertain about their affinities, although he suspected, wrongly, that many of them were remains of marine animals. He fully recognised the importance of comparative anatomy for a correct understanding of fossil vertebrates (Guettard, 1768, p. 15): 'But these comparisons will remain very incomplete until anatomy comes to the rescue. Only then will he [the naturalist] be able to determine accurately to which animals the bones he finds in the earth belong . . .'

Guettard's cautious attitude toward the identification of fossil vertebrates did not prevent him from giving the first illustration of a North American fossil mammal, in a monograph on the supposed geological resemblances between Switzerland and Canada, published in 1756. The specimen was a tooth of the American mastodon, which Guettard did not try to identify. Such fossil remains of large extinct mammals from North America were to play a major part in the debate about the meaning of vertebrate fossils during the second half of the eighteenth century (Simpson, 1942).

The first important discovery was apparently made by a French Canadian officer, Baron Charles de Longueuil (1687–1755). In 1739 Longueuil, a major in the French army, left Montreal with French and Indian troops to go to the aid of the governor of New Orleans, who was then fighting the Chickasaw Indians. While going down the Ohio River, they came across a number of large bones on the edge of a marsh. They thought that they were the remains of three elephants and collected some of them. In 1740, after the successful conclusion of the war, Longueuil took the fossils to New Orleans and then to Paris, where they were placed in the King's collection in the Jardin des Plantes. The tooth illustrated by Guettard in 1756 was apparently one of those collected by Longueuil (Simpson, 1942).

Longueuil's discovery was soon followed by other finds in the same general area. As noted by Simpson (1942), English settlers in the region of the Ohio River in what is now Kentucky, already knew about large fossil bones in the 1740s. One locality in particular was to become famous under the name of Big Bone Lick (from the salty quality of the bone-bearing mud, which attracted numerous wild animals). Much material was found there and sent to England, especially by George Croghan, an Irish adventurer who collected fossils when he was not fighting the Indians. In 1767 Croghan sent a collection of Big Bone Lick fossils to London, part of it going to Lord Shelburne (who was in charge of the American colonies), and the remainder to Benjamin Franklin.

The two main problems now were to identify this animal and to determine whether it was extinct or was still lurking in the depths of unexplored American forests. Guettard, in his memoir of 1756, did not propose any identification. In 1764 another French naturalist, Louis-Jean-Marie Daubenton (1716–1800), tried to solve the problem by comparing a femur brought to France by Longueuil with that of a modern elephant and that of a Siberian mammoth. This was a nice exercise in comparative anatomy, and Daubenton's conclusion was that all three femora belonged to elephants. Tusks found near the Ohio River confirmed the occurrence of elephants, but the molars also found by Longueuil and subsequent explorers posed a more complicated problem, as they were quite different from those of living elephants. The nearest teeth Daubenton could find to these mastodon molars were those of the hippopotamus. He finally concluded that the teeth were those of a large hippopotamus, and had not really been associated with the tusks and bones.

Benjamin Franklin thought that the remains from Big Bone Lick were those of elephants, and reminiscent of similar finds from Peru. In a letter written to Croghan in 1767 (quoted by Simpson, 1942), he noted that the tusks agreed with those of elephants, but that the 'grinders' were different, and reminiscent of those of a carnivorous animal. He went on to remark that 'it is remarkable, that elephants now inhabit naturally only hot countries where there is no winter, and yet these remains are found in a winter country'. Such peculiar occurrences of elephant bones had already been reported from Siberia, an even colder country. Franklin was thus led to the interesting suggestion that it looked 'as if the earth had anciently been in another position, and the climates differently placed from what they are at present'. In 1768, in a letter to a French correspondent (quoted by Simpson, 1942, p. 146), he changed his mind about the significance of the molars from Big Bone Lick: the idea of a carnivorous elephant-like creature seemed unlikely to him, because such an animal would have been 'too bulky to have the Activity necessary for pursuing and taking Prey'. He rather thought that the peculiar knobs on the teeth from Big Bone Lick were 'only a small variety', that is, a minor variation on the usual morphology of elephant teeth. He remarked that such variations were known to occur in many species, and that such knobs could be useful to grind twigs and branches.

In 1768 the English naturalist Collinson published two papers on the Big Bone Lick fossils. He too was puzzled by the occurrence of an elephant-like animal in a region with severe winters. As to the nature of the mysterious animal, he came close to the truth when he suggested that the tusks and molars belonged to the same form, which could be either a new species of elephant, or some completely unknown animal. He believed that the animal from Big Bone Lick was a herbivore, but there is no evidence that he thought it was extinct (Simpson, 1942).

At about the same time, the famous British anatomist William Hunter was also studying the Croghan collection. In a paper published in 1769, he announced that the teeth had been examined by ivory workers, who declared that they were 'perfectly similar' to those of living elephants. Nevertheless, Hunter came to the conclusion that this only showed that true ivory could be produced by two different animals. A comparison of a lower jaw collected by Croghan with that of an elephant convinced Hunter that the American form belonged to a different species, which he variously called a 'pseudelephant', an *animal incognitum*, or the 'American *incognitum*'. This remarkable beast, according to Hunter, had been carnivorous (Hunter, 1769): 'And if this animal was indeed carnivorous, which I believe cannot be doubted, though we may as philosophers regret it, as men we cannot but thank Heaven that its whole generation is probably extinct.'

Although Hunter's speculations about the diet of the 'American *incognitum*' were erroneous, he was certainly a precursor in suggesting that the animal was probably extinct. It took a long time for this startling idea to become generally accepted. Even skilful comparative anatomists who had noticed that fossil mammals were in some respects different from their closest living counterparts, found it difficult to admit that they could belong to 'lost' species. J.F. Merck (1741–91), for instance, who studied the bones of elephants and rhinoceroses from many German localities, thought it possible to relate these remains to species still living in Africa or Asia. In the same way, he thought that a crocodilian skull from the Liassic of Altdorf was similar to that of the living gavial of India (Tobien, 1984).

While French and British explorers were bringing to light the remains of the 'American *incognitum*', Russian expansion into northern Asia was revealing a wealth of equally surprising fossil vertebrates. As mentioned above, the abundant mammoth remains of Siberia had been known for a long time, not only to the native tribes, but also to the Chinese. At the end of the seventeenth century, European travellers came back from adventurous journeys into Siberia with stories of whole mammoth carcasses preserved in the frozen ground (Pfizenmayer, 1926; Abel, 1939a). In 1722 Baron Kagg, a Swedish officer who had been a prisoner of war in Siberia, came back to Sweden with a fantastic 'portrait' of the mysterious mammoth. This picture, provided by a Russian, showed a rather ox-like quadruped with powerful claws and a pair of spirally twisted horns. Also in 1722, Czar Peter I (the Great) ordered that all large bones found in Siberia should be excavated and sent to the capital; rewards were to be paid to the discoverers. What these animals were remained a mystery: when the Czar sent a Dr Messerschmidt to Siberia to investigate a carcass found in 1724 on the banks of the Indigirka River, this naturalist came back with a report of pieces of skin covered with long hairs, and the conclusion that this animal was none other than the Biblical Behemoth.

As more discoveries were made during the eighteenth century, however, it gradually became clear that the Siberian mammoth was also some kind of elephant. As mentioned above, as early as 1764 Daubenton had shown the close skeletal resemblances between the mammoth, the American mastodon and the elephant.

The identification of the mammoth as an elephant posed an obvious problem: how could carcasses of tropical animals such as elephants be found in the frozen ground of such cold countries as Siberia? As indicated earlier, the same question had been raised by several of the naturalists who had studied the elephant-like 'American *incognitum*'. One of the first scientists who tried to answer this question in a synthetic, comprehensive way was Buffon.

Georges Louis Leclerc, Comte de Buffon (1707–88), was undoubtedly the most famous scientist of eighteenth-century France. For many years he was the intendant of the Jardin du Roi, and this position gave him access to one of the largest natural history collections available at the time. His great *Histoire naturelle* (Natural history), written with several collaborators (including Daubenton) and published in many volumes between 1749 and 1789, was extremely popular. Although much of Buffon's zoology was rather anthropomorphic and more suited to the public taste than to the requisites of scientific description, the more speculative parts contain interesting insights on the variability of species, and suggest that Buffon came close to the idea of evolution. However, it is in the geological parts of his *Histoire naturelle* that he mentioned fossil vertebrates. In the *Théorie de la terre* (Theory of the earth), published in 1749, he noted that for some fossils, no living analogues had yet been found. This could be explained in two ways:

● in the case of marine forms, such as ammonites, living representatives may still remain undiscovered in the depths of the ocean;
● in the case of large terrestrial animals, there was a distinct possibility that some species had 'perished'. It was not possible to relate the extraordinary fossil bones found from Canada to Siberia to any known living animal.

What could have caused the extinction of such animals was not clear. Buffon was no diluvialist. He accepted the Biblical story of the Flood, but in contrast to Scheuchzer, he thought the Deluge had nothing to do with the origin of mountains or the accumulation of marine shells at high elevations. To him, the Flood was a miraculous event which had had little influence on the general history of the earth. He even anticipated nineteenth-century uniformitarianism when he wrote (Buffon, 1749) that

> causes the effect of which is unusual, violent and sudden should not bother us; they do not occur in the ordinary march of nature; however, effects which occur everyday, movements which succeed each other and are renewed without interruption, constant and always reiterated operations, these are our causes and our reasons.

The consequence of this view was that Buffon explained most of the changes which have occurred at the surface of the globe by the action of the sea and of erosion by running water. There was no need for large-scale catastrophes.

In 1778 Buffon published the *Epoques de la nature* (Epochs of nature), a supplement to the *Histoire naturelle* which was a complete revision of his *Théorie de la terre*. Fossil remains of large mammals were listed among the 'monuments' which can be regarded as witnesses of the first ages of the earth. They could be used, in conjunction with experimental evidence and traditional stories about the beginnings of the world, to reconstruct the early history of the earth. Using these different kinds of evidence, Buffon proposed a rather grandiose view of the several 'epochs of nature'.

The first epoch was when the earth and the other planets were formed (possibly by the collision of a comet with the sun). The earth was then in a state of fusion.

The second epoch was that of the solidification of the earth.

During the third epoch, the sea covered the whole surface of the globe, and limestones were formed by the accumulation of seashells.

The fourth epoch was marked by the retreat of the sea from the present continents and by the beginning of volcanic activity.

During the fifth epoch, elephants, hippopotamuses and other 'southern' animals inhabited the northern regions.

The sixth epoch saw the separation of the present continents.

Finally, the seventh epoch was defined as that 'when the power of man fecundated that of nature'.

The basic idea behind this rather revolutionary view of earth history was that our planet had originally been in a molten state, and had gradually cooled down. Within this framework (based on experiments with heated metal spheres), fossil vertebrates were an important piece of evidence, as they were supposed to show that large animals which today are restricted to the tropics had once lived in northern regions which are now too cold for them. The occurrence of these fossils was an extraordinary fact, which however could not be doubted. They were found not only in Siberia, Russia and western Europe, but also in North America, as shown by the discovery of the animal from the Ohio River. It could no longer be claimed, as had been done in the past, that so many large exotic animals had been brought to the northern regions by man. They really had lived where their remains were found, and this showed that these cold regions had once been as warm as our present torrid zone. The constitution of these animals could not have changed sufficiently 'to give the elephant the temperament of the reindeer'. The naturalist Gmelin, who had travelled in Siberia, had supposed that great floods in southern regions had forced the elephants to migrate toward the northern countries, where the cold had killed them. Buffon could not accept this interpretation: there were too many bones and tusks of these animals in Siberia, and there was no reason why the elephants should have migrated northward rather than eastward or westward.

Most of these large mammals whose bones were found in such unexpected places, had living respresentatives. The tusks found in Siberia and North America were elephant tusks, not those of walruses. There were hippopotamus teeth too. Under the influence of Daubenton's work on fossil 'elephants', Buffon had reconsidered the opinion he had expressed in the *Théorie de la terre*; in an 'addition' published in 1778 together with the *Epoques de la nature*, he remarked (p. 432):

> Among terrestrial animals, I know only one lost species; it is that of the animal whose molar teeth I had drawn with their dimensions in the *Epoques de la nature*; the large teeth and big bones which I could obtain have belonged to elephants and hippopotamuses.

The tooth in question was that of an American mastodon. Buffon realised that such teeth could not have belonged to true elephants. He believed that they indicated a huge animal, an 'ancient species which must be regarded as the first and largest of all terrestrial animals', and which 'existed only in early times and has not persisted until us' (such a large animal could not have remained undetected until the present day). Its closest relative seemed to be the hippopotamus.

The former presence of these large 'tropical' animals in northern countries was easy enough to explain: during the early stages of earth history, the earth's own heat was much greater than that it received from the sun. Elephants could thus live in northern regions. When the surface of the globe became cooler, the climate near the poles became colder, and the elephants had to migrate southward. Geographical displacement, however, did not cause morphological changes. Fossil elephants were not different from the living ones; they may have been larger, however, simply because 'Nature was in its first vigour', so that giant forms could appear. Size decrease came later, as the earth cooled down. As to the extinct species, both terrestrial and marine, such as the Ohio animal and the ammonites, they must have been adapted to higher temperatures than those of today's torrid zone. They had become extinct when the climate had become too cold for them everywhere.

Buffon's original view of earth history implied that long periods of time had elapsed before man's appearance. He thought that 6,000 to 8,000 years had elapsed since Adam, but the study of God's work in nature showed that the earth was much older, possibly as much as 75,000 years old (in unpublished manuscripts, Buffon even considered a 3 million-year-old earth, with life appearing about one million years ago; see Albritton, 1980). Buffon had made estimates about the age of the earth, based on his experiments on cooling metal spheres. He thought that some 35,000 years must have elapsed after the formation of the globe before temperatures became tolerable and life could appear. The first living species must have been less sensitive to heat than the later ones, and they had become extinct when the climate had become too cold for them. Terrestrial animals appeared only later, probably in the cooler northern regions, about 15,000 years ago. Buffon thus was the originator of the idea of a successive appearance of various forms of life on earth, as the outside conditions gradually became suitable for them.

The occurrence of abundant remains of elephants and other large mammals in many northern countries had another unexpected implication: it showed that Europe, Asia and North America had once been contiguous before they became separated during the sixth epoch (this separation, Buffon insisted, had nothing to do with the Flood). This led Buffon to make some interesting remarks on what was later to be called palaeobiogeography. He discussed possible migration routes for large mammals between Eurasia and North America, and came to the conclusion that a connection across the Bering Strait was more likely than a transatlantic route. He had also noticed that South America had a distinct fauna of its own, and he thought that this relatively isolated region had 'produced' smaller and weaker animals than the other parts of the world, and had produced them later. He also speculated that elephants had become extinct in North America because they had been prevented from migrating southward by the high mountains of Central America.

Buffon's conception of earth history was revolutionary indeed: not only was the earth much older than was previously thought, but living beings had appeared gradually over long periods of time, before man's final appearance. Some species had even become extinct because of environmental changes. Although Buffon claimed that his aim was to 'reconcile forever natural science and theology', some theologians did not like his rather heretical interpretation of the 'epochs of Nature'. In 1751 some propositions in the *Théorie de la terre* and other volumes of the

Histoire naturelle had already been denounced as contrary to the beliefs of the Church by theologians of the Sorbonne, and Buffon had had to publish a not very convincing recantation. The *Epoques de la nature* also attracted the attention of the censors, but Buffon enjoyed the king's protection and there was no condemnation (Gascar, 1983).

Theologians were not alone in opposing some of Buffon's theories. There were scientific opponents as well, among whom was the German naturalist Peter Simon Pallas (1741–1811), who had been invited to work for the St Petersburg Academy of Sciences in 1767 by the Empress Catherine II, and had taken part in a long expedition to Siberia from 1768 to 1774. In a communication to the St Petersburg Academy published in 1777 and later as a separate volume in 1779, Pallas presented 'observations on the formation of mountains and the changes which happened to the globe, to serve the natural history of M. le comte de Buffon'. In this work he drew on observations he had made in Siberia, where he had found, along the banks of most rivers, abundant remains of the 'large animals of India': elephants, rhinoceroses and monstrous buffaloes. One of his most striking discoveries had been that of a carcass of a rhinoceros, found with its skin in the frozen ground on the banks of the Viloui River. This, to Pallas (1779, p. 71), was 'convincing proof that it must have been a most violent and most rapid flood which once carried these carcasses toward our glacial climates, before corruption had time to destroy their soft parts'.

Pallas had first thought that northern countries had once had milder climates, but the sight of so many remains of large mammals had convinced him of the reality of a catastrophic flood. Pallas's theory was that submarine volcanoes had given birth to the islands of the East Indies and the Philippines. These considerable movements of the sea-floor had expelled a huge quantity of water, which had engulfed the mountain chains of Asia and Europe, and had carried the remains of drowned animals from India to their northern flanks; the valleys of Siberia, in which these remains were now found, had been carved by the force of this gigantic flood. Pallas went on to equate this tremendous catastrophe with the floods of the Chaldaean, Persian, Indian, Tibetan and Chinese mythologies, and with the Biblical Deluge.

This concept was obviously very different from the gradual processes acting over long time spans envisaged by Buffon. In many ways, Pallas's hypothesis announced the catastrophist theories of the early nineteenth century.

Unlike Pallas, other naturalists accepted the extinction of some species, but they offered alternative explanations to Buffon's 'climatic' hypothesis. The 'overkill' theory already used by Molyneux to explain the *local* extinction of the Irish 'moose' was resorted to by the French naturalist Defay to explain the *complete* disappearance of other species. In a collection of observations on the local natural history of the Orléans region published in 1783, Defay described several fossil bones and teeth which had been found in a limestone quarry at Montabuzard, a fossil locality now known to be Miocene in age. He identified some of the remains as those of a deer, and also mentioned a tooth of a hippopotamus much larger than the living ones (this was probably a worn mastodon tooth). Two incomplete molars were attributed to Buffon's 'only terrestrial animal whose species is lost', because they were similar to a figure of a mastodon tooth published in the *Epoques de la nature*. Defay was puzzled by the occurrence in the same deposit of remains of animals which he thought could not have lived together. To explain this, he envisaged a 'very considerable flood' which had gathered together the remains of animals from different regions. He accepted Buffon's explanation of the remarkably large size of fossil animals and of the subsequent 'degradation' of species. On the subject of extinctions, however, he had ideas of his own. He tried to account for the extinction of the enormous quadruped mentioned above in the following way (Defay, 1783, p. 63):

It is indeed very conceivable that this monstrous species (I call it so only on account of its extraordinary size) must have succumbed under the blows of hunters or of packs of large carnivores; because its prodigious mass must have been a hindrance to its rapidity of movement; as it lacked the necessary agility to resist the attacks of so many enemies, it had to hide in the less frequented places; but could it find salvation there? Let us judge of the excessive size of an animal whose jaws must have borne sixteen molar teeth, each one weighing ten to twelve pounds, and it will be understood that there was no refuge where the vigilant eye of the hunter could not discover it; and even if we suppose that it could have taken shelter in a cavity of some rock, it would have sufficed, to find it, to follow the deep tracks its enormous mass must have left on the ground.

Defay had no doubt about the reality of extinctions. In his own words, 'it is thus in the nature of things that some species become lost'. He speculated on the coming extinction of some living animals, such as the South American sloths, which were hunted by man for their flesh and had no means of escaping their fate. Future generations would possibly encounter the fossil remains of sloths, just as we sometimes find those of several extinct species of marine or terrestrial creatures.

Nevertheless, at the end of the eighteenth century there was no consensus about extinct species, and some naturalists still thought that the living representatives of the so-called 'lost' species would sooner or later be found by explorers in remote parts of the world (this was, for instance, the opinion of Thomas Jefferson, second president of the United States). The final demonstration of the reality of extinction was still to come, and every important find of fossil bones excited renewed speculations, all the more so because interest in fossils was becoming increasingly widespread.

This growing interest was to result in early attempts at the scientific excavation of fossil localities. Thus, in a book published in 1802, the Italian scientist Alberto Fortis published a description of excavations which had been conducted eighteen years before near Verona. In his article on 'fossil elephant bones from Romagnano, in the Verona region', he recounted how his friend Count Gazola had found a heap of elephant bones on one of his country estates at a place called Romagnano. As Fortis thought that 'fossil objects become important for the advancement of science only when a good account can be given of their matrix and of the accompanying circumstances', he made several visits to the site and made detailed observations. The bones had been found by a farmer when ploughing a field. Fortis and Gazola had the topsoil removed, and thus obtained a 49 square-foot (4.9m^2) area, which was excavated. A thin layer of yellowish clay contained large bone fragments and limestone blocks. The bones had been broken and scattered before fossilisation, and they were fused to limestone blocks and lumps of iron oxide by a calcareous stalactite. Teeth and bones of small mammals and birds were found in the deposit, but neither Gazola nor Fortis could identify them for lack of comparative material (Fortis was sorry about this, as he recognised the potential importance of these microvertebrates). The dominant species at Romagnano, however, was the elephant. Several molars of young and adult individuals were unearthed, together with a large tusk, possibly 12 to 14 feet (3.5–4m) long, which was unfortunately broken by careless workmen. Fortis commented on the large size of the tusk, much bigger than those of living elephants. He could not help feeling relieved that such large elephants, which must have been capable of fearful devastations, no longer existed.

The main problem about the Romagnano fossil locality was the complete absence of evidence concerning its antiquity. Fortis realised that the number of elephants brought into Italy by the Ancients for military purposes or for circus games was not sufficient to account for the numerous fossil remains which had been unearthed in that country. In an early attempt at taphonomy, he distinguished several kinds of fossil localities: in some instances, articulated bones were found in bogs and marshes; in others, scattered bones were discovered in marine deposits; in a third kind of locality there was evidence of human activity, the bones having been accumulated by man. This third type could be subdivided into those 'artificial deposits' in which tusks were present, and those where no ivory occurred. Fortis thought that this distinction was of great importance, as 'the presence or absence of ivory indicated very different epochs'. He had no doubt that the Romagnano site belonged to the third type: the elephant bones and tusks had been buried there by men, and this on several occasions. The place where the bones were found was much too small for several elephants to have died there from natural causes, and there was no evidence of fluviatile or marine activity. As Fortis did not believe in great floods or similar extraordinary physical events, the only explanation left was that the elephants had been buried by men who were not civilised enough to know the value of ivory. The bones and tusks had probably been buried for religious reasons, as part of more or less superstitious rites (this was reminiscent of Guettard's ideas concerning the fossils from Etampes). The savages who thus buried elephant remains with their tusks in the foothills of the Alps must have lived many centuries ago, but there was no need to accept the extremely long time-scale postulated by Buffon's theory of the cooling earth. Interestingly, Fortis thought that elephants did not necessarily require a very warm climate to survive. He also rejected diluvialist explanations. Pallas's hypothesis was unacceptable to him, because the Siberian remains did not show evidence of transportation over a long distance.

Finally, Fortis proposed a chronology of elephant localities:

• the oldest remains were found in marine beds, and they dated back to a time when the sea covered almost the whole surface of the globe. The fossils from the Arno valley in Tuscany belonged to this period, and they were those of enormous marine animals, which had later become terrestrial when the sea receded;
• more or less articulated and complete skeletons with their tusks were those of animals which had accidentally died in marshes;
• broken remains with tusks had been buried by savages; the Romagnano locality was a good example. Fortis suggested that it possibly was at least 15,000 to 20,000 years old;
• elephant remains found without tusks belonged to the historical period, when men knew the value of ivory and had removed the valuable tusks.

Fortis's report on the Romagnano finds illustrates the situation of vertebrate palaeontology (a term not yet in use at the time) at the end of the eighteenth century: although interest in fossil vertebrates was great enough to lead to systematic excavations, the precise identification of the bones still posed major problems. Comparisons with living animals were rather superficial, and the existence of completely extinct species was not yet clearly recognised by all naturalists. It was becoming obvious that all fossil vertebrates were not equally old, but their chronological succession was far from established. On all of these especially important points rapid progress was to be made at the turn of the nineteenth century.

4

Blumenbach, Cuvier and Earth's Revolutions

More than anyone else, Georges Cuvier contributed to the final demonstration of the reality of extinction. Some of his contemporaries, however, had come to very similar conclusions about the fossil record. Certainly, the most important of them was Johann Friedrich Blumenbach (1752–1840). Blumenbach had been a student of J.E. Walch, from whom he derived a lasting interest in fossils. He became a professor at the University of Göttingen, and a large part of his research work was devoted to comparative anatomy and anthropology. As early as 1799, he distinguished three main groups of fossils in his *Handbuch der Naturgeschichte*:

- the *petrificata superstitorum* were easily identifiable, because they were perfectly similar to extant organisms; many animals and plants from the famous Oeningen locality belonged in this category;
- the *petrificata dubiorum* resembled extant living beings, but they were distinct from them because of their enormous size, or because of various small anatomical differences; creatures such as the cave bear, *Ursus spelaeus*, or the extinct elephant, *Elephas primigenius*, were typical representatives of these 'dubious' fossils;
- the *petrificata incognitorum* were fossils of completely unknown creatures of the past, for which no living analogues could be found; the mastodon, which Blumenbach called *Mammut ohioticum* (he was one of the first to apply Linnaean names to fossils) was one of them, as well as the ammonites and the belemnites.

In later works Blumenbach was to arrange this classification according to a chronological order, and to relate it to what in 1799 he had already called 'the more or less general catastrophes which have affected our earth'. He insisted on the importance of accurate comparisons between fossil and living forms, and of careful study of the 'geognostic' relations between different fossil localities.

In 1803 the description of a few invertebrate fossils from Hanover gave him the opportunity to express his views about what he called the 'archaeology of the earth' (they were expressed again in a more general book in 1806). A first class of fossils contained remains of animals and plants identical to the living ones and similar to those now living in the same region. These remains were the most recent of all. The Oeningen fossils were supposed to belong to that class.

A second class comprised remains of creatures which had died during the Flood; they were often incomplete because of the action of the waters, and included exotic but still-extant forms which had been transported from distant places. A few remains indicated forms totally different from the living ones: the Ohio *incognitum* was one of them.

A third and older class was created for the fossils which indicated universal changes of the earth's climate. This corresponded to the *petrificata dubiorum* of 1799: these fossils were slightly different from today's exotic animals. The numerous bones of elephants and rhinoceroses found in Germany belonged to this class, as well as the bones of bears, lions and hyaenas from German caves (these bears, found together with tropical beasts, must have been adapted to a warm climate). The tropical animals found in the lithographic limestones of Pappenheim and Eichstätt, in Bavaria, were also placed there. Among them were horseshoe crabs, small starfish, and remains of a large bat similar to the flying fox (this must have been a pterodactyl). The fossils of this class usually showed no signs of long transportation: they had died where they had lived. Their extinction was due to 'a total change of climate', the causes of which were unknown.

The fourth class contained the oldest fossils, those which showed that the surface of the primitive earth had been largely covered by water. Ammonites which were sometimes found on high mountains were a good example of that class: strong crustal movements had brought such fossils to high altitudes. They had probably been subjected to multiple revolutions, but it was not yet possible to subdivide these ancient events.

According to Blumenbach, a parallel could be drawn between these chronological classes of fossils and the 'ages' of classical tradition: the recent fossils of the first and second classes corresponded to the 'historical' age, those of the third class to the 'heroic' age, and the oldest remains of the fourth class to the 'mythical' age.

Not only was Blumenbach convinced of the reality of extinctions, he also believed that they affected whole faunas, not just a few individual species. As he wrote in 1806 (p. 13), it was 'more than likely that it is not only one genus or another which has become extinct on our earth, but a whole organised preadamite creation'. There had been 'total revolutions', and the 'ages', each of them characterised by a particular 'class' of fossils, were separated by such 'revolutions'. The first class of fossils, those which were identical with living forms, had been formed after the last general 'revolution', during merely local catastrophes. The last 'total revolution' had been the Flood.

Thus, at the turn of the nineteenth century, Blumenbach had understood that a large number of species had become extinct, that these extinctions had not all been simultaneous, and that successive faunas, markedly different from each other, had peopled the earth in the course of geological time. The history, or 'archaeology', of the earth could be reconstructed by establishing the succession of these faunas and of the catastrophes which had destroyed them. This was a considerable conceptual advance, despite the fact that Blumenbach's chronology was largely wrong. However, most of the extinct species were still only vaguely identified, and their place in the zoological classification uncertain. Cuvier's work was to provide a much better understanding of the composition of extinct faunas and of their succession.

Georges Cuvier (1769–1832) was born in Montbéliard, then a French-speaking possession of the Duke of Württemberg. He became interested in natural history at an early age, and this interest increased during his studies at the Caroline Academy in Stuttgart. He moved to France in 1788, and spent most of the French Revolution in Normandy, where he worked as a tutor in a noble family and had the opportunity to conduct anatomical researches on various groups of invertebrates. He came to Paris in 1795, and soon obtained a position in the Comparative

Anatomy Department of the newly founded Muséum d'Histoire Naturelle. He soon rose to a prominent position not only in the Museum but also in the whole educational system of the French Empire (for an interesting new assessment of Cuvier's life and career, see Outram, 1984; older biographies by French authors tend to be either over-critical or hagiographic). Political changes after the fall of Napoleon in 1815 little affected his career, and until his death he occupied various high administrative positions.

Cuvier's interest in invertebrate comparative anatomy and classification never completely disappeared, but soon after his arrival in Paris he became interested in fossil vertebrates, and started to publish important contributions on this subject. As early as 1796, he described an unusual fossil mammal, the so-called 'animal from Paraguay' (1796a). The rather complex story of the discovery and description of this remarkable specimen is worth telling in some detail. This was in fact the skeleton of a Pleistocene giant ground sloth, found in 1788 by the Dominican friar Manuel Torres near Luxan (or Luján), 65 km (40 miles) west of Buenos Aires, in what was to become Argentina. The specimen was sent to Madrid by the viceroy of Buenos Aires, and it arrived at the Royal Collection of Natural History in September 1788. It was the object of much curiosity there, and it is even reported that King Charles III of Spain gave orders to the effect that a living specimen should be procured, or, if that proved impossible, a stuffed one (Ingenieros, 1957). The skeleton was prepared and mounted by J.B. Brú, a preparator for the Royal Collection. This giant sloth thus became the first fossil vertebrate to be displayed in a more or less life-like pose (Simpson, 1984). Brú also wrote a description and had drawings made and engraved, but neither the description nor the illustrations were immediately published. Foreign scientists also became interested in the strange animal: at the end of 1793, the Dane P.C. Abildgaard saw the skeleton in Madrid, and published a short, poorly illustrated description of it in 1796 (in which he compared it with the South American anteater and the aardvark). In 1795 the Frenchman Philippe-Rose Roume, an official of the French Government in Saint-Domingue, visited Madrid on his way to France and managed to obtain a copy of Brú's plates. He sent them to the French Institute in Paris, possibly with Brú's approval (Hoffstetter, 1959). Roume then published a short note on the skeleton in which he mentioned Cuvier's opinion that the animal showed affinities with the sloths. The plates were passed on to Cuvier, who was asked to prepare a report on them for the Institute. This report was published as a short illustrated paper in 1796 (1796a). Cuvier recognised that the giant animal from Luján was related to the sloths, and he gave it a Linnaean name, *Megatherium americanum* (this was rather untypical of Cuvier, who rarely bothered to give proper Linnaean names to the fossils he described — which later resulted in much taxonomical confusion when several names were proposed for the same animal by different translators or subsequent authors). A Spanish engineer by the name of Joseph Garriga then heard about Cuvier's work on the strange animal whose skeleton was kept in Madrid. He convinced Brú to sell him his plates and his unpublished description, and finally published a paper in 1796 which included Brú's description and plates, as well as a translation of Cuvier's note. The whole complicated story has been interpreted in a way rather unfavourable to Cuvier, who has been accused of unethical behaviour by several authors, among them Simpson (1984), but as shown by Hoffstetter's detailed analysis of the case (Hoffstetter, 1959), Cuvier apparently did not know who had prepared the plates, and had not heard about Brú's description, and the charges against him thus seem to be unfounded.

In the same year, 1796, Cuvier read a paper before the French Institute on the living and fossil species of elephants. In this paper (1796b) he first showed that the Asian and African elephants were more different from each other than the horse from the donkey or the goat from the sheep. He next tackled the problem of fossil elephants. His interpretation of the remains from Europe, Siberia, North America and Peru was of some importance to geology, which he defined as a science which tried to reconstruct the 'revolutions' the globe has gone through. Most of the hypotheses which had been put forward to explain the occurrence of fossil elephants in cold regions (floods, transportation by man, Buffon's cooling earth) were in fact unnecessary, because detailed anatomical study of these remains showed that they were not close enough to living elephants to be included in the same species. The teeth and jaws of the Siberian mammoth were not completely similar to those of modern elephants, and the Ohio animal was even more different. Cuvier again used a simple comparison to make this point clear to a fairly wide public: these animals were as different from each other as the dog, the jackal and the hyaena. There was thus no reason to assume that the mammoth and the Ohio animal had had the same mode of life as our elephants; they had probably been able to live in a cold climate. However, this interpretation led into new difficulties; what had happened to these animals which no longer existed? Cuvier went on to list a number of fossil vertebrates which could be shown to differ from living species: the Siberian rhinoceros, the cave bear, the reptile from Maastricht, the newly found 'animal from Paraguay'. None of them had living representatives. Why, on the other hand, were there no true fossil men? All these incontrovertible facts seemed to 'prove the existence of a world anterior to ours, and destroyed by some catastrophe'. It now remained to reconstruct this primitive world and the catastrophe which had destroyed it. Comparative anatomy had revealed this new field of research, but it was left to other, more audacious 'philosophers' to explore it more fully. This conclusion to Cuvier's paper may have been purely rhetorical: in the years following, he was audacious enough to provide the first systematic identifications and reconstructions of the inhabitants of this former world.

Cuvier's early works were followed by a long succession of papers on various fossil vertebrates, in which he was able to explain the anatomical principles he applied to the reconstruction and identification of 'lost' species. A large amount of material was easily available to him from the vast gypsum quarries then being exploited in the vicinity of Paris. Abundant fossil remains of late Eocene vertebrates were being found there by the quarrymen, and several authors, including Guettard, had already speculated, without much success, about the affinities of these animals. Cuvier's strict scientific approach was to yield spectacular results, despite great difficulties. As he remarked in his *Discours sur les révolutions de la surface du globe* (Discourse on the revolutions of the surface of the globe), first published in 1812 as an introduction to his *Recherches sur les ossemens fossiles*, complete skeletons of quadrupeds were rare; usually the naturalist had to work on a jumble of isolated and broken bones. The bones from the gypsum of Montmartre and other places around Paris were abundant but usually much fragmented, both by natural processes and by the methods used by the quarrymen to extract them. Moreover, as Cuvier soon recognised, they were the remains of extinct animals only distantly reminiscent of living forms, so that simple comparison with the latter could give no satisfactory clues as to their identification. Only a careful application of the principles of comparative anatomy could provide the necessary guidelines, and the most essential of these principles was that of the 'correlation of forms in organised beings'. In his *Discours*, Cuvier enunciated it as follows (1834, t. 1, p. 178):

> Every organised being forms a whole, a unique and closed system, the parts of which correspond to each other and contribute to the same definitive action by a reciprocal reaction. None of these parts can change without changes also occurring in the others; consequently, each of them, taken separately, indicates and gives all the others.

Simple examples could be given: if, for instance, the intestines of an animal were so organised that it could only digest meat, then its jaws had to be able to swallow its prey, its claws to seize and tear it apart, its teeth to cut and divide it; its limbs had to enable it to pursue and overcome its prey, its sensory organs to see it from afar, and its brain to provide it with the instincts it needed to stalk and ambush its victims. These correlations could be followed into the minutest details, and the shape of every part of the body could give accurate information about the systematic position of the animal. Cuvier (1834, t. 1, p. 181) even used mathematical analogues to explain his method:

> In a word, the shape of the tooth implies the shape of the condyle, that of the scapula, that of the nails, just as the equation of a curve implies all its properties; and just as by taking each property separately as the basis of a particular equation, one would find again both the ordinary equation and all the other properties, similarly the nails, the scapula, the condyle, the femur, and all the other bones taken separately, give the tooth, or give each other; by beginning with any one of them, someone with a rational knowledge of the laws of organic economy could reconstruct the whole animal.

Cuvier applied this method to the bones from the gypsum quarries. He has given a vivid description of the problem he was faced with, and of his feelings when he discovered he could solve it:

I was in the situation of a man to whom would have been given the jumbled, mutilated and incomplete remains of several hundred skeletons belonging to twenty kinds of animals; each bone had to be returned to those with which it was properly associated; it was almost a small-scale resurrection, and I did not possess the all-powerful trumpet; but the immutable laws prescribed to living beings replaced it, and, when comparative anatomy spoke, each bone, each fragment of bone, went to its proper place. I have no words to depict the pleasure I felt when I saw how, each time I discovered a character, all the more or less predicted consequences of this character successively developed; the feet were in conformity with what the teeth had announced; the teeth with what the feet announced; the bones of the legs, the thighs, all those which had to connect these extreme parts, turned out to be shaped as could be predicted; in a word, each species was reborn, so to speak, from a single one of its elements.

In 1804 Cuvier published a first monograph on 'the animal species from which come the fossil bones scattered in the gypsum of the Paris region' (1804a). He had recognised that some parts were more useful than others for identification purposes; 'the first thing to do,' he wrote, 'is to recognise the shape of its molar teeth'. In this way he was able to quickly discover that most of the animals from the gypsum quarries had the molars of herbivorous 'pachyderms'. On the basis of several jaws and incomplete skulls, he was able to reconstruct the head of what he thought was a tapir-like animal, for which he coined the name *Palaeotherium medium*. This was followed by a second monograph on the remains of other animals, among which was a larger form, about the size of a cow, which he called *Palaeotherium magnum*, a small *Palaeotherium*, several species of a new genus, *Anoplotherium*, a carnivore, and some reptiles and birds. Over the years the list of fossil vertebrates identified by Cuvier from the gypsum quarries steadily grew longer, until a whole extinct fauna could be reconstructed. One of its most unexpected components was probably the famous 'sarigue de Cuvier' (Cuvier's opossum), which gave him the opportunity for a brilliant demonstration of his anatomical method (Cuvier, 1804b). On the basis of the dentition, he had recognised that a small skeleton from Montmartre belonged to a marsupial, and therefore predicted that marsupial bones would be found when the pelvic region was completely prepared. When he removed the matrix, the marsupial bones duly appeared.

Cuvier was also the first to publish scientific reconstructions of fossil animals as they may have appeared in life. In his great work *Recherches sur les ossemens fossiles* (first published in 1812, fourth and last edition 1834–36), he published a plate, drawn by his collaborator Laurillard, showing the reconstructions of two species of *Palaeotherium* and two species of *Anoplotherium*; these were widely reproduced by subsequent authors in both scientific and popular books.

From the beginning, Cuvier tried to extend his researches to as many fossil vertebrates as possible. In the early 1800s he sent a circular letter to a large number of naturalists in various European countries in which he requested information about fossil bones. The political circumstances of the time helped him in this project: much of continental Europe was then under French occupation, and Cuvier's high position in the educational system of the French Empire gave him the opportunity to travel to several countries (notably Italy and the Netherlands) where he could study collections of fossil vertebrates. He was thus able to examine Scheuchzer's *Homo diluvii testis* in the Teyler Museum in Haarlem, and to establish that it was a giant salamander. His enquiry about fossil bones had been successful, and he received a great deal of information from various foreign correspondents (although his relations with some of them sometimes became somewhat strained, as the exchange of information occasionally tended to be one-sided; for an instance concerning the Dutch naturalist Van Marum, see Touret, 1984).

Although the bulk of the fossils which Cuvier had the opportunity to study were those of Tertiary and Pleistocene mammals, he also successfully applied his methods to reptilian bones from older formations. As early as 1800 he had studied bones unearthed from the Jurassic strata of Normandy by local collectors (Buffetaut, 1983; Taquet, 1984), and had identified them as those of gavial-like crocodiles. In his 1808 work on these remains, he also illustrated and described dinosaur vertebrae, which he thought belonged to an unusual type of crocodile (Taquet, 1984).

In 1801 Cuvier first mentioned the skeleton of a small reptile, apparently able to fly, which had been found in the lithographic limestones of the Eichstätt region of Bavaria. The fossil had become part of the natural history collection of Karl Theodor,

Elector of the Palatinate and Bavaria, in Mannheim, sometime between 1767 and 1784 (Wellnhofer, 1984), and had first been described by Cosimo Alessandro Collini in 1784. Collini (1727–1806), an Italian from Florence, had been Voltaire's secretary before becoming Karl Theodor's private secretary and director of the Mannheim natural history collection. Although he understood the importance of comparison with living animals, Collini could not find any modern counterpart for the fossil, with its long-toothed jaws and peculiarly elongated forelimbs. He thought it was related neither to birds nor to bats, and finally concluded that it must have been some kind of marine animal (Wellnhofer, 1984).

Cuvier's attention had been drawn to this remarkable specimen by a Strasbourg professor, Johann Hermann, who thought it could be a mammal intermediate between bats and birds, but Cuvier did not endorse this opinion. He had to base his identification on Collini's excellent illustration, as he could not locate the specimen (which had been transferred to Munich after Karl Theodor had moved his residence from Mannheim to the Bavarian capital). His conclusion was that the animal was a reptile, and that it was able to fly thanks to the wing-membrane which presumably attached to the enormously elongated fourth finger. For this remarkable reptile, Cuvier coined the name 'Ptero-Dactyle', and he reconstructed its mode of life as that of an active insectivorous flyer. He later had to defend this interpretation against that of the Bavarian anatomist Samuel Thomas von Soemmerring, who had the fossil in Munich and had described it in 1812 as a bat-like mammal which he had named *Ornithocephalus antiquus*. Although Cuvier's opinion was based on much sounder anatomical comparison and was rapidly accepted by most scientists, the idea that pterosaurs may have been somehow related to mammals was still accepted by some authors until the 1840s (see Wellnhofer, 1984, for a detailed account of the interpretation of the first pterosaur).

Another Mesozoic reptile which was successfully reinterpreted by Cuvier was the famous 'Great Animal of Maastricht'. The fossil, a very large skull, had been found in late Cretaceous strata in the vast underground quarries of Saint Peter's Mountain, near Maastricht, in 1780. A local military surgeon and fossil collector by the name of Hofmann had arranged for the specimen to be carefully chiselled out of the surrounding limestone. However, the owner of the ground, a Canon Godin, claimed that the skull belonged to him, and filed a lawsuit against Hofmann, who lost his case. The fossil had to be returned to Godin, who kept it in his house in Maastricht. In 1795 the city was besieged by the French revolutionary army under General Pichegru. The French geologist Faujas de Saint-Fond, who was then scientific commissioner of the Republic for the Netherlands, knew about the fossil skull, and arranged for Godin's house to be spared when the city was bombarded. Godin, however, became suspicious about this unexpected favour, and concealed the specimen in a cellar. When French troops took the city, the fossil at first could not be found, but after a reward of six hundred bottles of wine was offered, a group of soldiers quickly located it and turned it over to the French authorities, who had it transferred to the Paris Museum, where it still is. Similarly, when the French army occupied Verona a few years later, Count Gazola was induced to 'donate' a large part of his collection of Eocene fishes from the famous Monte Bolca locality to the Paris Museum (Blot, 1969). Fossils were obviously considered as valuable plunder. Incidentally, it seems that this kind of behaviour persisted for a long time in the French army: according to a story told to the author a few years ago, in 1945 a famous French palaeontologist who was then a serving officer, arrived with a party of soldiers at the famous Hauff Museum in Holzmaden (south-western Germany) and 'expressed the wish' to obtain a rare ichthyosaur skeleton for the Paris collection — upon which the specimen was taken off the wall and removed to Paris.

The Great Animal of Maastricht had already attracted much attention when Cuvier became interested in it. In 1786 the Dutch anatomist Peter Camper had identified other similar bones from Maastricht as those of some kind of whale, an opinion shared by Van Marum. Hofmann thought it was a crocodile, and Faujas de Saint-Fond, in the large volume he published on the natural history of Saint Peter's Mountain in 1799, endorsed this opinion and tried to support it with a review of fossil crocodiles. In 1799 Peter Camper's son Adrian concluded that the mysterious animal from Maastricht was neither a whale nor a crocodile but some kind of giant lizard (Camper, 1799). Cuvier's comparative studies fully confirmed Adrian Camper's conclusion: the Great Animal of Maastricht was indeed closely related to the lizards; it was an enormous monitor-like reptile, as large as a crocodile, and to judge from the rocks in which it was found, it must have been marine. Cuvier did not give it a scientific name; the British palaeontologist Conybeare later called it *Mosasaurus*.

When in 1812 Cuvier published a collection of his previously published works on fossil vertebrates under the title *Recherches sur les ossemens fossiles*, he had already described a fairly large number of mammals from various Tertiary and Pleistocene strata, as well as a few strange Mesozoic reptiles such as the pterodactyl and the mosasaur. By comparison with the rather superficial descriptions and uncertain identifications of most eighteenth-century authors, this was tremendous progress, and it had important consequences. First of all, it was becoming increasingly difficult, not to say impossible, to deny that many species had become extinct. Creatures such as the pterodactyl or the *Megatherium* could not be confused with known living animals. It could still be claimed that the living counterparts of the animals reconstructed by Cuvier would eventually be found alive in still unexplored parts of the world, and to counter this argument Cuvier asserted in his *Discours* that there was little hope of discovering hitherto unknown large quadrupeds. To support this assertion, he tried to show that the terrestrial fauna of all parts of the world was now well known. That this was a hasty generalisation was shown shortly thereafter when two young French naturalists who had studied under Cuvier, Duvaucel and Diard, reported the discovery of an hitherto unknown tapir in the Malay Peninsula. This totally unexpected find of a large living mammal was followed throughout the nineteenth century by a long succession of similar discoveries of unknown animals, ranging from the gorilla to the giant panda (Heuvelmans, 1955). Nevertheless, Cuvier's view was not totally wrong: it remained a fact that no living counterparts could ever be found for the vast majority of fossil vertebrate species. The reality of extinction could no longer be doubted.

A second important point which emerged from Cuvier's research was that all fossils were not equally old: several faunas had succeeded each other on earth. Before Cuvier, several naturalists, including Blumenbach, had realised this, but Cuvier went much farther in his reconstruction of the succession of extinct faunal assemblages. To do so, he turned to the study of the stratigraphy of the fossil-bearing rocks, especially those around Paris which had yielded so many remains of *Palaeotherium*, *Anoplotherium* and other extinct mammals. To elucidate the relationships between the various sedimentary strata of the Paris Basin, Cuvier obtained the help of a young geologist, Alexandre Brongniart. As reported by Cuvier in the *Recherches sur les ossemens fossiles*, for four years they went together on weekly excursions, drew profiles of quarries, measured sections, compared fossil-bearing strata over long distances. Mineralogical analyses were performed by Brongniart, who certainly did most of the work, as Cuvier himself acknowledged. The result was the *Essai sur la géographie minéralogique des environs de Paris* (Essay on the mineralogical geography of the Paris region), first published in an abridged form in 1811, and which became a very long chapter of the *Recherches sur les ossemens fossiles*. Cuvier and Brongniart were able to reconstruct a sequence of seven main geological formations, from the chalk at the bottom to the alluvial gravels at the top. A coloured 'geognostic' map summarised their findings. Their detailed study of the depositional environment of these formations revealed alternating marine and freshwater beds, and the ossiferous strata could be placed with some accuracy within this general framework. The different formations were distinguished not only by their lithological characters, but also by the fossils found in them. This, of course, was not a completely new approach. Several eighteenth-century authors, among whom Giraud-Soulavie (who had studied the geology of parts of south-eastern France), had already recognised the importance of fossils for a chronological subdivision of sedimentary rocks. As early as 1799 the British surveyor William Smith, working along roughly the same lines, had prepared a manuscript geological map of the surroundings of Bath (Faul and Faul, 1983), and although his first printed map was not published until 1815, his efforts in this direction antedated those of Cuvier and Brongniart by a few years. Interestingly, Smith and the Cuvier-Brongniart team arrived at roughly the same results from rather different starting points: while Smith was involved in what would now be called 'economic geology', Cuvier and Brongniart were interested in stratigraphy and geological mapping in a more purely intellectual way — which may simply reflect the more advanced economic and industrial stage reached by Britain at that time.

Be that as it may, the geological researches on the Paris Basin contributed to Cuvier's attempt at a synthetic explanation of the succession of extinct faunas he had discovered. This was presented in the *Discours sur les révolutions de la surface du globe*, first published as an introduction to the *Recherches sur les ossemens fossiles*, and later issued as a successful separate volume which went through many editions, even after Cuvier's death. This *Discours* has been considered as the most controversial part of Cuvier's work. The main problem was to explain how the successive faunas had become extinct, before being replaced by a new assemblage. Cuvier's answer was that the earth had suffered several 'revolutions' which had brought about major faunal changes. Rudwick (1972) has rightly remarked on Cuvier's use of the term 'revolution' a few years after the end of the French Revolution, but Cuvier was definitely not the first naturalist to understand earth history in terms of violent episodes. Pallas and Blumenbach, to mention but two, had already made use of such catastrophes to explain accumulations of fossil bones or faunal changes. Cuvier accepted the then current idea that the accidents of the earth's surface demonstrated the occurrence of considerable and violent upheavals: folded and faulted strata were obvious evidence of such movements, which had sometimes brought sediments formed under the sea to high altitudes. Moreover, the alternating freshwater and marine strata of the Paris Basin showed that several of these upheavals had taken place in the course of geological time. Finally, these violent events had been sudden; proof of this was afforded by the frozen carcasses of large quadrupeds found in Siberia, which must have been frozen very soon after death before their soft parts had rotted away (this was Pallas's old argument, and contradicted Cuvier's earlier views about the possible adaptation of mammoths to a cold climate). Cuvier's conclusion was that 'life has often been troubled on this earth by frightful events. Numberless living beings have fallen victim to these catastrophes' (Cuvier, 1834, vol. 1, p. 108). An important point was that present geological processes were unable to produce such tremendous catastrophes; in Cuvier's words (1834, vol. 1, p. 117), 'the thread of operations is broken, the march of nature has changed'.

Those 'frightful events' had caused the disappearance of whole faunas. As remarked by Laurent (1984), Cuvier's fossil species were not only 'lost', they were 'destroyed', and he had come to this conclusion in the 1790s, before he became acquainted with the geology of the Paris Basin, probably under the influence of German naturalists such as Blumenbach. The mechanisms which Cuvier envisaged for such destructions were not quite clear. The 'broken thread' referred to above suggests radical changes and the complete destruction of all life (or at least all terrestrial life) on earth. However, Cuvier apparently did not think that such 'revolutions' had destroyed all living beings — which would have implied several divine creations to re-people the earth after each complete destruction. In several parts of the *Discours*, he mentions 'partial revolutions'. How he envisaged faunal change at the time of a revolution is explained by a famous example: if a great flood was to cover New Holland (Australia), it would completely wipe out the various marsupials which inhabit this continent and are not found elsewhere. If the same 'revolution' turned the several straits separating Australia from the Asian mainland into dry land, migration routes would become available to Asian mammals such as elephants, rhinoceroses, camels, tigers and buffaloes, which would then colonise Australia (after the retreat of the flood waters). A naturalist would then discover that the fossil fauna of Australia was completely different from the living one. If a second 'revolution' was to destroy Asia, it would be extremely difficult to determine the origin of the new Australian fauna. This kind of explanation was largely based on negative evidence, and could hardly provide an explanation for the ultimate origin of the various extinct faunas. Nevertheless, Cuvier also applied it to man. No human bones had ever been found among fossil vertebrates; supposed human fossils had turned out to be misinterpreted animal remains, or recent bones to which a high antiquity had erroneously been assigned. Contrary to what has often been written, Cuvier did not deny the possible existence of fossil man. He thought that before the last catastrophe man did not inhabit the regions where fossil remains of terrestrial animals had been found (i.e. most of Europe, Asia and America), but may have lived in restricted areas of the globe where he had been able to survive the last 'revolution' and to re-populate the earth afterward. This last 'revolution' had taken place a relatively short time ago (it could not be much more than 6,000 or 5,000 years), and Cuvier devoted many pages of his *Discours* to a discussion of historical, archaeological and astronomical data (Brach, 1984) to demonstrate this. Obviously, there were records of this 'revolution' in the traditions of many countries, which all mentioned a catastrophic flood. Interestingly, Cuvier did not especially try to equate this particular event of earth history with the Biblical Flood in particular. Despite his involvement in Protestant organisations, he apparently was not the devout Protestant depicted by many historians. According to Outram (1984, p. 145), remarks made by his friends and relatives indicate that 'his personal belief consisted of a minimal deism', and he certainly was not a 'Biblical geologist' who tried to interpret palaeontological data in scriptural terms. Whatever this last 'revolution' was, it had been preceded by several earlier ones; at least two or three 'irruptions of the sea' had taken place in Europe, and their reality was demonstrated by faunal replacements.

Cuvier's discoveries exerted a considerable attraction on the public imagination. The demonstration that the world had once been peopled by wholly extinct faunas, which consisted of such astonishing creatures as flying reptiles and giant lizards, was hailed as a major conquest of the human mind. Although some aspects of Cuvier's personality and social behaviour were resented by his colleagues, his well-written books (especially the *Discours sur les révolutions de la surface du globe*) quickly became popular, and he was acclaimed by many as an immortal genius. In his novel *La peau de chagrin* (1831), for instance, Honoré de Balzac grew enthusiastic:

> Have you ever dived into the immensity of space and time, by reading the geological works of Cuvier? Carried by his genius, have you floated above the limitless abyss of the past, as if you were carried by the hand of a wizard? When discovering, layer by layer, bed by bed, under the quarries of Montmartre or in the schists of the Urals, these animals whose fossil remains belong to antediluvian civilisations, the soul is frightened to perceive billions of years, millions of peoples which the weak human memory, the indestructible divine tradition, have forgotten, and whose ashes, deposited on the surface of our globe, form the two feet of soil which give us bread and flowers. Is not Cuvier the greatest poet of our century? Lord Byron reproduced in words a few moral agitations; but our immortal naturalist has reconstructed worlds from withered bones . . .

The response to Cuvier's work in scientific circles was somewhat different. The use he had made of anatomical principles to reconstruct extinct vertebrates was generally accepted and admired, but his interpretation of the fossil record was severely criticised by some naturalists who favoured an evolutionary approach rather than his vision of repeated destructions. Jean-Baptiste Lamarck (1744–1829), the most famous of early nineteenth-century evolutionists, had studied the fossil invertebrates of the Paris Basin, and was more impressed by continuity than by discontinuity. There was no room for extinct species in his theory of evolution, as he proposed it in his *Philosophie zoologique* (1809). A general progressive trend carried all living beings toward higher levels of organisation, and secondary mechanisms such as use and disuse and the efforts of individuals to adapt to their environment, coupled with the inheritance of acquired characteristics, explained the diversification of the living world. The individual was the only biological reality, and species were inconstant and variable. However, changes at the specific level were so slow that they could not be noticed by man. Lamarck devoted a short chapter of his book to 'lost species'. It seemed unlikely to him that the means used by nature to ensure the preservation of species could be so inadequate as to allow the destruction of whole 'races'. Nevertheless, the fossil record did reveal a multitude of animals, among which very few had perfectly identical living counterparts. Could it be concluded that such species no longer existed? Lamarck found it difficult to accept such an idea. He noted that there were still many unexplored regions (including the depths of the ocean) where supposedly extinct species might still be found alive. Another possibility was that man was responsible for the extermination of some species (Lamarck, 1809, p. 76):

> If there are really lost species, this doubtless can only be among the large animals which live on dry land, where man, by the absolute empire he exerts, could succeed in destroying all the individuals of some of those he did not want to preserve or to domesticate. Hence the possibility that the animals of the genera *palaeotherium*, *anoplotherium*, *megalonix*, *megatherium*, *mastodon* of Mr Cuvier, and a few other species of already known genera, no longer exist in nature: nevertheless, this is a mere possibility.

However, extermination by man could explain only the extinction of large terrestrial animals. The aquatic, and especially the marine, species, and the very small terrestrial animals, were so numerous and so difficult to capture that they could not be totally destroyed in this way. Nevertheless, many fossil molluscs were known which had no living analogues. As man could not be held responsible for their disappearance, how could they belong to really extinct species? The likeliest explanation was that such fossils belonged to still-extant species which had changed through time, so that their living representatives were now somewhat different from the fossil ones. It was all the more likely that nothing on earth was constant. Therefore, it could be 'felt' that living beings also had to vary because of changes in their environment and habits. The really astonishing fact about fossils was that a few of them had living counterparts. These were probably the least ancient of all fossils, which had not had time enough to change significantly since their appearance. There was thus no need to assume that 'universal catastrophes' were responsible for the extinction of species (Lamarck, 1809, p. 80):

> It is to be regretted that this easy way out, when one wants to explain the operations of nature, the causes of which are unknown, has no foundation, except in the imagination which created it, and cannot be supported by any proof.

There were indeed local catastrophes (such as earthquakes and volcanic eruptions), but gradual change was the rule, and all operations of nature could be explained by it.

Lamarck's remarks were obviously aimed at Cuvier, and expressed strong opposition to his catastrophist doctrine. Cuvier's long discussion of the unlikelihood of future discoveries of large quadrupeds in the *Discours* may have been intended as an answer to Lamarck's criticisms. Moreover, Cuvier emphatically denied any close genealogical links between fossil and living species, and attacked 'those who believe in the indefinite possibility of alteration in the shapes of organised bodies, and think that . . . all species can change into each other' (Cuvier, 1812). If such transformations had occurred, intermediate fossil forms should have been found, but there was no evidence of 'so curious a genealogy', for the simple reason that extinct species had been just as constant as living ones. For Cuvier, the fixity of species was a perfectly established fact, and he thought that he had demonstrated it through his studies of animal mummies (notably those of the ibis) brought back from Egypt by Bonaparte's expedition: these 5,000-year-old creatures were no different from living species, and this showed that no evolution had taken place. In the absence of any absolute time-scale, it could still be claimed that 5,000 years represented a very considerable time-span, a not insignificant part of geological time (although, as we have seen, much longer durations had already been suggested by some eighteenth-century authors for the evolution of the earth).

There was not much scientific debate on the question of extinct vertebrates between Cuvier and Lamarck. As remarked by Outram (1984, p. 154), 'they talked past each other'. There was little in common betweeen Lamarck's speculations, which were still reminiscent of eighteenth-century philosophy, and Cuvier's much more factual approach to the events in the history of life. Cuvier's position on the origin of species was ambiguous; the mechanism he envisaged for the repopulation of the earth after each of his major 'revolutions' was unclear. He certainly did not believe in multiple divine creation after each catastrophe, but neither could he accept Lamarck's evolutionary interpretation.

A more direct and detailed alternative to Cuvier's interpretation of fossil vertebrates was provided in the 1820s and 1830s by his colleague at the Natural History Museum, Etienne Geoffroy Saint-Hilaire (1772–1844). Geoffroy Saint-Hilaire had been among those who urged Cuvier to move to Paris in 1795, and they had collaborated on zoological works in 1795 and 1796. Their conceptions soon diverged, however, and they gradually became scientific adversaries. Cuvier could not accept Geoffroy's ideas about the unity of plan in animals, which ran contrary to his own ideas about anatomy and classification. In 1825 Geoffroy Saint-Hilaire intruded on Cuvier's favourite field of investigation, fossil vertebrates. The opportunity for this was provided by new discoveries of fossil crocodilians in the Jurassic rocks of Normandy. In 1817 an incomplete skeleton had been found in a limestone quarry near Caen. Parts of it were given to a medical student named Luard by a quarryman who had been his patient (Eudes-Deslongchamps, 1896). Another student, who was to become a prominent vertebrate palaeontologist, J.A. Eudes-Deslongchamps, understood the interest of the remains and, with some friends, managed to obtain more fossils. Eventually, some of them were sent to Cuvier, who also obtained parts of a second crocodilian from another quarry near Caen (Buffetaut, 1983). Cuvier realised that the crocodiles from Normandy were different from the living ones in some respects, but he arbitrarily associated fragments belonging to different species, thus creating what Eudes-Deslongchamps later called 'anatomical monsters'.

Geoffroy Saint-Hilaire soon grew interested in the new fossils from Caen. In 1825 he showed that they were generically distinct from the living gavial, and created the genus *Teleosaurus* for the specimens from Caen, while the fossils from Le Havre and Honfleur described much earlier by Cuvier were placed in another new genus, *Steneosaurus*. Geoffroy Saint-Hilaire went far beyond the mere taxonomic allocation of these fossils by trying to interpret them in evolutionary terms. In his 1825 publication (pp. 152–53), he wrote:

It is not repugnant to reason, that is to physiological principles, that the crocodiles of the present epoch could be descended through an uninterrupted succession from the antediluvian species found today in the fossil state on our territory.

This seems to be the earliest interpretation of specified fossil vertebrates as the direct ancestors of living forms. Following some of Lamarck's ideas, Geoffroy Saint-Hilaire explained this evolution by the influence of environmental changes. He went even further, and suggested that, whereas *Steneosaurus* was more primitive than living crocodiles, *Teleosaurus* was more advanced, and could be considered as a kind of intermediate link between crocodiles and mammals. As he expressed it (1825, p. 144), 'to consider *Teleosaurus* as if it was a mixture of mammal and crocodile is to conceive its essential and true affinities'. Although his interpretations were fanciful, Geoffroy Saint-Hilaire was obviously much ahead of his time; such clearly evolutionary interpretations of fossil vertebrates (and fossils in general) would remain exceedingly rare until evolution became accepted by a large part of the scientific community after the publication of Darwin's *Origin of Species*.

Geoffroy Saint-Hilaire thought that the fossil reptiles from Normandy provided excellent material to demonstrate his ideas on evolution. He exchanged an abundant correspondence with Eudes-Deslongchamps, who meanwhile had become professor of natural history at the University of Caen. In 1830 and 1831, Geoffroy Saint-Hilaire travelled to Caen (Bourdier, 1969) to examine newly found fossils, which became the basis of several publications (Geoffroy Saint-Hilaire, 1831). In these, he again proposed evolutionary interpretations: he now saw *Teleosaurus* as an intermediate between ichthyosaurs and modern crocodiles.

He also tried to reconstruct the appearance of teleosaurs, and came to the strange conclusion that they had large fleshy lips, or even a proboscis, and fed on algae. For some reason, he also believed that their feet bore a large median toe with small rudimentary digits on the sides.

Geoffroy Saint-Hilaire later became increasingly interested in theoretical and philosophical questions, and fossil vertebrates were used mainly as a support for his sometimes rather abstruse speculations (which Eudes-Deslongchamps, apparently disappointed, called 'lucubrations'). When Eudes-Deslongchamps found an incomplete dinosaur skeléton, Geoffroy Saint-Hilaire again travelled to Caen, in 1837, to see it, and was enthusiastic about the specimen (Bourdier, 1969). Fossil mammals also attracted his attention, although he published relatively little on them. In a note on Tertiary mammals from central France (1833), among which he had found intermediate forms, he gave his opinion about the appearance of new living beings, which he thought was a gradual process induced by environmental changes. Later, in 1837, he engaged in a controversy with the anatomist Blainville on the affinities of *Sivatherium*, a fossil giraffe which had just been discovered in the Siwaliks of India by Falconer and Cautley. This was again the opportunity for him to attack Cuvier's ideas on the immutability of species.

Strangely enough, Cuvier did not react to Geoffroy Saint-Hilaire's interpretation of the crocodiles from Normandy. In the last edition of the *Recherches sur les ossemens fossiles*, published posthumously between 1834 and 1836, there is no mention of Geoffroy Saint-Hilaire's memoirs on these fossils. When a fierce controversy between Cuvier and Geoffroy Saint-Hilaire broke out at the Academy of Sciences in 1830, it revolved around the unity of plan among the major groups of animals, and although the mutability of species was mentioned, fossil vertebrates apparently played very little part in the debate.

Geoffroy Saint-Hilaire's conceptions found some favour among a part of the public who did not like Cuvier's rather conservative ideas (see, for instance, Gosse's satirical 'natural history' of the professors of the Jardin des Plantes, published in 1847, in which Cuvier's ideas and social behaviour are sharply criticised, whereas Geoffroy Saint-Hilaire and Lamarck are given as models to all naturalists). Nevertheless, they seem to have exerted little immediate and direct influence on most of the scientific community. One reason for this may have been that

Geoffroy Saint-Hilaire's written and spoken style was long-winded and boring. His writings are in an abstruse pseudo-philosophical jargon which makes them difficult to read. It is reported (Bourdier, 1969) that during a session of the Academy of Sciences in 1834, most of the members, including Geoffroy Saint-Hilaire's own son Isidore, left the room when he embarked on a ponderous lecture on the influence of the environment on plant evolution! By contrast, Cuvier's style was elegant and clear — hence the popularity of some of his works (notably the *Discours sur les révolutions de la surface du globe*) even outside scientific circles.

Be that as it may, the general scientific atmosphere in the 1820s and 1830s seems to have been unreceptive to Geoffroy Saint-Hilaire's ideas. Vertebrate palaeontology developed considerably during this period, but it did so mainly along Cuvierian lines.

5

Early Nineteenth-century Controversies

Cuvier's works had revealed a series of extinct faunas which had succeeded each other through geological time. As mentioned earlier, this rather sensational discovery, which thoroughly modified our understanding of earth history, much excited the imagination, and it is no wonder that interest in palaeontology (the word was coined by Blainville about 1825) grew enormously among both scientists and laymen at the beginning of the nineteenth century. The result of this increased interest was, not unexpectedly, the discovery of many new fossil localities which revealed an unsuspected array of strange, extinct creatures.

Most vertebrate fossils were still obtained, as they had previously been, from quarrymen or others who spent much time on rocky outcrops because of their occupations. Obtaining fossil vertebrates from quarries could be time-consuming and frustrating. J.A. Eudes-Deslongchamps has related (Eudes-Deslongchamps, 1838) how in the 1830s he managed to recover a fairly large part of a skeleton of a large reptile (now known to have been a carnivorous dinosaur) which had been found in a limestone quarry in the vicinity of Caen. The first remains of the animal were noticed in a limestone block on a building site by a friend of Eudes-Deslongchamps. Before the latter could visit the site, however, street urchins had already hammered some of the bones out of the stone, and he had to hunt for the missing parts in the neighbourhood. He finally managed to determine in which quarry the fossil had been found, but a visit there did not yield any result. A few days later, however, a quarryman brought him several fragments he had extracted himself — without taking the slightest precautions. Finally, three weeks later, another quarryman provided other fragments, and indicated where another limestone block with more remains of the same animal could be found. From what must have been a fairly complete skeleton, only a mass of broken bones and a few fragments in large limestone blocks remained. It took Eudes-Deslongchamps much time and effort to reassemble them as well as could be done in the circumstances (Eudes-Deslongchamps, 1838; Buffetaut, 1983).

As interest in them developed, fossils quickly acquired some commercial value, and high prices were sometimes paid by collectors for especially interesting specimens. Enterprising individuals who had the good fortune to live close to a fossil locality soon understood that collecting and selling fossils could be a profitable business. The most famous early nineteenth-century example of this was that of the Anning family of Lyme Regis on the Dorset coast (Howe, Sharpe and Torrens, 1981). Richard Anning had been selling fossils to visitors to this seaside resort to supplement his income, and when he died in 1810, his children Joseph (1796–1849) and Mary (1799–1847) continued the family business. They became famous for their discovery in 1811–12 of an ichthyosaur skeleton in the lower Liassic strata exposed along the coast. The skeleton, which was to be described by Sir Everard Home in 1814, was purchased from the Annings for £23. Mary Anning's fame as a fossil collector subequently grew when she discovered plesiosaur and pterosaur remains. On the other side of the English Channel, fossil collecting had also become a part-time job for some inhabitants of the coastal towns of Normandy: in 1828, for instance, A. de Caumont, in an essay on the geology of Calvados, noted that a fisherman from Villers-sur-Mer by the name of Bloche, had been collecting and selling fossils for many years.

An idea of the high prices sometimes fetched by fossil vertebrates is given by a short paper on the price of fossils published by Gideon Mantell in 1846. In 1820, for instance, an ichthyosaur specimen collected by Mary Anning was auctioned for £152. 5s. (£152.25). Later, the first *Plesiosaurus* skeleton, also found by Mary Anning, was purchased by the Duke of Buckingham for 150 guineas (£157.50). Incomplete specimens were of course cheaper: an ichthyosaur femur was purchased for Cuvier at the 1820 auction for £1. 10s. (£1.50). Fossils also were imported from abroad: Mantell mentions that in 1836 isolated mastodon teeth from Big Bone Lick were sold for between £2 and £4. 4s. (£4.20); the British Museum, however, had to pay 140 guineas (£147) for an incomplete skull.

Although public museums did acquire fossil vertebrates, many of the specimens went to private collections. There were few professional palaeontologists, and it was still possible for amateurs with a sufficient background in natural history to make valuable contributions not only to the discovery, but also to the scientific study of fossil vertebrates. The most famous of these early nineteenth-century amateur palaeontologists certainly (and deservedly) is Gideon Algernon Mantell (1790–1852). Mantell had started to collect fossils around his native town of Lewes, in Sussex, when he was still a schoolboy. He eventually became a surgeon and started to practise medicine in Lewes in 1818. Fossil collecting soon came to occupy a large part of his time and activity, and after he had discovered dinosaur remains, his main interest became fossil vertebrates. Mantell corresponded with the leading geologists of his time both in England and abroad, and he was the author of many papers and books on geology and palaeontology. His passion for fossils was eventually to make his last years unhappy (Colbert, 1968). After moving to Brighton in 1833 he gradually turned his house into a museum, and relations with his family became strained. In 1839 his wife, who at first had been his collaborator in the field of palaeontology, left him. In 1844 Mantell moved to London, and finally sold his large palaeontological collection, which he said had cost nearly £7,000 to assemble, to the British Museum for £4,000. Although he had never occupied an official scientific position in a university or museum, Mantell was undoubtedly one of the most active British palaeontologists of his time, and the part he played in the discovery of dinosaurs was considerable.

On the continent too, some leading palaeontologists never held official scientific positions. One of them was Hermann von Meyer (1801–69). He came from an old Frankfurt family, and studied finance and natural history before he entered the financial administration of the Bundestag (the German parliament) in 1837, and became director there in 1863. His considerable palaeontological researches thus had to be conducted in his spare time. In addition, he was the founder of *Palaeontographica*, one of the first journals solely devoted to palaeontology. Interestingly enough, in 1860 von Meyer, who had become a leading authority on fossil reptiles, turned down an appointment as professor at the University of Göttingen, because he felt that his position in the financial administration better enabled him to preserve his independence (Tobien, 1974).

One of the leading French vertebrate palaeontologists of the first half of the nineteenth century, Edouard Lartet (1801–71), also began as an amateur. Lartet had studied law and had become a lawyer in his native region of south-western France. He often gave free advice to the local farmers, and one of them once brought him a fossil tooth as a present. This mastodon tooth was apparently the initial cause of Lartet's interest in palaeontology, which led him to explore the Miocene vertebrate localities of his native region. Among his major finds were the remains of Miocene apes (*Pliopithecus* and *Dryopithecus*), which shook the current idea that the higher primates, like man, were newcomers which had appeared only during the latest period of earth history. Lartet later became interested in early man, and his excavations in the caves of south-western France greatly contributed to the recognition of the existence of fossil man by the scientific community. In 1869 he was appointed to the prestigious position of Professor of Palaeontology at the Paris National Museum. One often overlooked achievement of Lartet was that he was among the first palaeontologists to conduct systematic excavations of fossil localities. As mentioned in Chapter 1, there had been crude excavations at Cannstatt in 1700, and in the 1780s Fortis and Gazola had excavated the Romagnano elephant locality, but most palaeontological collecting was still largely a matter of chance discovery: fossils were picked up from or dug out of sediments during geological excursions, or they were bought from quarrymen or specialised dealers. Lartet, however, was not content with the isolated fossils which local farmers could provide. In 1834 a shepherd directed him to a hill near the small village of Sansan where fossil bones had been found. He started excavating at once, and was able to publish a preliminary note about what was to become a major Miocene locality, in 1835. The Minister of Public Instruction was interested, and funds were made available for further research (this may well be the first instance of state-funded palaeontological excavations, if one excepts the early dig at Cannstatt), which took place in 1837 and 1838. The fossils thus collected were sent to the Paris Museum. In 1847 the locality was finally purchased by the French government and turned over to the National Museum (Lartet, 1851). Lartet had trenches dug into the fossil-bearing strata where they cropped out at the periphery of the hill, and between 15,000 and 20,000 cubic metres (18,000–22,000 cubic yards) were thus excavated. Besides the abundant bones of large mammals, remains of small vertebrates were found in a marly layer which also contained many crushed freshwater shells. As the small bones were not easy to collect in that kind of deposit, Lartet devised a simple method to obtain them: after the sediment had been dried, it was carefully washed and screened. The residue was then examined and the smallest bones could easily be picked out. This simple and efficient method for the recovery of microvertebrates was not to come into general use until the second half of the twentieth century.

Another category of amateur which developed in the first half of the nineteenth century was the clergyman-palaeontologist. Among both the Protestant and Catholic clergy there were many who did not feel that palaeontological discoveries contradicted divine revelation. God's will was manifested in His works as well as in His word, and the study of nature helped to understand the plan of Creation. Moreover, catastrophist theories were relatively easy to reconcile with the Biblical story of the Flood. Characteristically, for instance, the village priest in Gustave Flaubert's novel *Bouvard et Pécuchet* (published posthumously in 1884) encourages his parishioners to study geology because it proves the reality of the Deluge. William Buckland (1784–1856), who was a clergyman but by no means an amateur in the field of palaeontology, and eventually became Dean of Westminster (Rupke, 1983), certainly felt that fossil evidence supported the idea of a universal deluge. He interpreted his discoveries in a Pleistocene bone-bearing cave deposit in Yorkshire in this light, although his efforts to reconcile Science and Scripture were considered by some as a distortion of both (Rudwick, 1972). Less important clergymen than Buckland were also active collectors on a local scale. A good example is the Abbé Croizet, vicar of a small village in the Massif Central of France, who eagerly collected fossils from the Plio-Pleistocene deposits around his parish, and dedicated to Georges Cuvier the volume on fossil bones which he published with his collaborator Jobert in 1828. For Croizet, there was no contradiction between Cuvier's views and the Biblical story of Genesis.

With the assistance of a large number of enthusiastic amateur collectors, the palaeontologists of the early nineteenth century were able considerably to augment the number of fossil species of vertebrates which had been known to Cuvier when he published the first edition of his *Recherches sur les ossemens fossiles* in 1812.

As shown in Chapter 4, most of the fossils Cuvier had worked on until then were remains of Tertiary and Pleistocene mammals (although the pterodactyl and the mosasaur were notable exceptions). An important discovery of the early nineteenth-century followers of Cuvier, especially in England, was that before the epoch of the extinct mammals the world had been dominated by even stranger large reptiles.

The first to be discovered were marine forms. As early as 1699 Edward Lhwyd had published an illustration of an ichthyosaur vertebra (Howe *et al.*, 1981). Subsequently, during the seventeenth century, ichthyosaur skeletons were found in the Liassic bituminous shales of the Holzmaden area of Württemberg and taken to the Stuttgart 'Gymnasium', some of them eventually finding their way to the Stuttgart palaeontological collection; one of them, collected in 1749 and still preserved in the State Natural History Museum in Stuttgart, even shows an embryo within the body-cavity of its mother (Adam, 1971). However, it was in the second decade of the nineteenth century that ichthyosaurs were first scientifically described in detail. Although the specimen found by Joseph and Mary Anning at Lyme Regis in 1811–12 was by no means the first one to be reported, its description by Everard Home in 1814 marked the true beginning of the scientific study of these extinct marine reptiles. Home first thought that the animal was a kind of fish. The discovery of additional specimens from the same general area eventually led him to change his opinion, and in 1818 he concluded that ichthyosaurs showed affinities with the duck-billed platypus in some of their bones, and that they had been air-breathing, swimming animals. Finally, in 1819, Home coined the name *Proteosaurus*; he now thought that this creature was somehow intermediate between fishes and lizards. However, this strange being had already been named *Ichthyosaurus* by Charles Konig of the British Museum in 1818. Henry De la Beche (1796–1855), who was to become the first director of the Geological Survey of Great Britain, then named three distinct species of *Ichthyosaurus* in 1819, and finally gave a more complete and accurate description in 1821 in a joint paper with W.D. Conybeare (1787–1857). Cuvier, who had acquired some remains at an auction held in London in 1820 and had a few from France, incorporated *Ichthyosaurus* in later editions of his *Recherches sur les ossemens fossiles*. In Germany, Georg Friedrich Jaeger, the director of the State Natural History collection in Stuttgart, described ichthyosaur remains from the

Boll region in 1824.

Plesiosaur remains were first described by Conybeare and De la Beche in their 1821 paper, but it was not until 1824 that an almost complete skeleton found by Mary Anning at Lyme Regis revealed that this animal had an extremely long, swan-like neck. The specimen, purchased by the Duke of Buckingham, was described by Conybeare under the name *Plesiosaurus* (close to lizards). Cuvier thought it was the strangest of all extinct animals.

The ichthyosaur and the plesiosaur soon became the most popular of fossil vertebrates. Reconstructions were attempted by several authors, the most striking being those published by Thomas Hawkins (1810–89), an eccentric and enthusiastic collector from Glastonbury in south-western England, who built up a large collection of ichthyosaurs and plesiosaurs from various localities in that region (Howe *et al.*, 1981). Hawkins published a first book, *Memoirs of Ichthyosauri and Plesiosauri* in 1834, in which quite a 'reasonable' reconstruction of ichthyosaurs and plesiosaurs in a Jurassic landscape was given. In 1840 he brought out a more eccentric volume, entitled *The Book of the Great Sea Dragons, Ichthyosauri and Plesiosauri, Gedolim Taninim of Moses. Extinct Monsters of the Ancient Earth*. The frontispiece of this book introduced a great classic of nineteenth-century palaeontological reconstruction, the fight between the plesiosaur and the ichthyosaur. This had been anticipated by De la Beche's famous cartoon *Duria antiquior* (ancient Dorset), drawn around 1831 (McCartney, 1977), in which various Jurassic marine reptiles engage in furious combat. Hawkins's reconstructed animals did look like dragons, and their fierce fighting in a tempestuous Jurassic ocean was fit to fire the imagination of romantic readers. The scene was made even more nightmarish by the presence of a few demon-like pterodactyls. This dramatic episode of former worlds was reproduced under various guises in countless popular palaeontology books during the rest of the nineteenth century (and later). The theme of the battle between prehistoric 'sea-dragons' proved so attractive that Jules-Verne used it in his *Voyage au centre de la terre* (1864), in which the explorers are the terrified witnesses of such a fight in a vast underground sea. Ichthyosaurs and plesiosaurs thus became the archetypes of the 'prehistoric monsters', the extinct dragon-like giant reptiles. They were to keep this position in the public mind until the discovery of complete skeletons of dinosaurs later in the century made these terrestrial creatures even more popular.

The first dinosaur bones to be described, however, were too incomplete to give an accurate idea of the appearance of these animals. As mentioned above, part of a dinosaur femur from Oxfordshire had been interpreted as a giant's bone by Robert Plot as early as the end of the seventeenth century. In 1699 a *Megalosaurus* tooth had been illustrated by Lhwyd (Delair and Sarjeant, 1975). In 1758 Joshua Platt reported the discovery of enormous vertebrae and a huge thigh-bone in the 'slate-stone' pits at Stonesfield, in Oxfordshire. These bones certainly belonged to dinosaurs. Among the fossil bones from Normandy which had been sent to Cuvier were dinosaur vertebrae, which he mistook for those of unusual crocodiles, and it seems that William Smith had come across *Iguanodon* bones in the course of his geological mapping (Norman, 1985). However, it was not until the 1820s that palaeontologists began to realise that giant terrestrial reptiles had once inhabited the earth, and because of the fragmentary character of the available fossils, it took a long time before a reasonably accurate picture of these creatures could be gained.

The first dinosaur bones to be scientifically described and named came from the so-called Stonesfield slate, which was quarried in underground pits to be used as roofing tiles. William Buckland, then Professor of Geology at Oxford, described them in 1824 in a paper published in the *Transactions of the Geological Society of London* under the title 'Notice on the Megalosaurus or great fossil lizard of Stonesfield'. The available material consisted of a portion of lower jaw, some vertebrae, an incomplete pelvis, a piece of shoulder-blade, and several bones from a hind limb. Buckland had received advice concerning these fossils from other experts, including Conybeare and Cuvier. The latter had seen the specimens during a visit to England in 1818, and Buckland had sent him copies of his paper and his plates before they were published (Taquet, 1984). Cuvier provided an estimate of the size of the animal; he thought it must have been some 40 feet (12m) long. This was a huge size indeed for a reptile, and the generic name chosen by Buckland alluded to this striking feature. Moreover, the affinities of this creature were not perfectly clear. The blade-like teeth were inserted in sockets as in crocodiles, but other features, notably in the shoulder girdle, were more reminiscent of lizards. The way of life of *Megalosaurus* was equally obscure. Cuvier thought it was a voracious marine reptile the size of a small whale.

In a letter written to Cuvier in 1824 (Taquet, 1984), Buckland mentioned that he had seen many bones of *Megalosaurus* from the 'Iron-sand formation' of Sussex, in Gideon Mantell's collection. The latter's discoveries in the Wealden beds of south-eastern England were to bring palaeontology a step closer to an understanding of the true nature of dinosaurs. Mantell had been collecting fossils, including large bones, in Sussex for several years when, in 1822, an unusual tooth was discovered by his wife. He exhibited the tooth, together with other fossils, at a meeting of the Geological Society, where Buckland, Conybeare and others saw it, but could only offer the suggestion that it belonged to a fish. Only Wollaston supported Mantell's idea that it may have belonged to an unknown herbivorous reptile (Mantell, 1851). Mantell then decided to consult the leading authority on fossil vertebrates (Mantell, 1851, p. 228):

> As my friend Mr (now Sir Charles) Lyell was about to visit Paris, I availed myself of the opportunity of submitting it to the examination of Baron Cuvier, with whom I had the high privilege of corresponding: and, to my astonishment, learned from my friend, that M. Cuvier, without hesitation, pronounced it to be an upper incisor of a Rhinoceros.

To complicate matters further, Mantell soon thereafter discovered some metacarpal bones in the quarry which had yielded the tooth. These also were sent to Cuvier, who at first sight thought that they belonged to a hippopotamus. Buckland now suspected that all these remains came not from the 'Iron-sand formation', but from the 'superficial diluvium' (i.e. the Pleistocene), and he warned Mantell of possible stratigraphic mistakes. Fortunately, additional remains were soon discovered (Mantell, 1851, p. 230):

> Other specimens, however, were soon procured by stimulating the diligent search of the workmen by suitable rewards, and at length teeth were obtained which displayed the serrated edges, the longitudinal ridges, and the entire form of the unused crown. I then forwarded specimens and drawings to Baron Cuvier, and repaired to London, and with the aid of that excellent man, the late Mr Clift, ransacked all the drawers in the Hunterian Museum that contained jaws and teeth of reptiles, but without finding any that threw light on the subject. Fortunately, Mr Samuel Stuchbury [sic], then a

young man, was present and proposed to show me the skeleton of an Iguana which he had prepared from a specimen that had long been immersed in spirits; and, to my great delight, I found that the minute teeth of that reptile bore a closer resemblance in their general form to the fossils from Tilgate Forest, than any others with which I was able to institute a comparison.

Meanwhile, in Paris, Cuvier had had second thoughts about the curious teeth submitted to him by Mantell. In June 1824 he wrote him a letter (reproduced in Mantell, 1851, and Taquet, 1984) in which he admitted that they probably belonged to reptiles, although their outward appearance was somewhat reminiscent of fish teeth:

> Could we not have here a new animal, a herbivorous reptile? and just as today in terrestrial mammals it is among the herbivores that the largest species are found, similarly, in the reptiles of the past, when they were the only terrestrial animals, could not the largest of them have fed on plants? A part of the large bones which you possess would belong to this animal, hitherto unique of its kind. Time will confirm or reject this idea . . .

In the second edition of the *Recherches sur les ossemens fossiles*, Cuvier admitted that his first impression had been wrong, and that both the large bones and the strange teeth found by Mantell could not be attributed to large mammals such as the hippopotamus or the rhinoceros, but in all likelihood belonged to a most unusual kind of giant reptile.

In 1825 Mantell finally published a description of this new fossil reptile, which was named *Iguanodon*, following a suggestion by Conybeare, 'to indicate the resemblance between the fossil teeth and those of the recent Iguana'. The available material was still rather scanty, and Mantell thought that his *Iguanodon* must have resembled the modern iguana in its general appearance, although it was much larger. In a famous sketch, he reconstructed the skeleton of *Iguanodon* as lizard-like, with a small horn on the snout (much later, the 'horn' was recognised as a thumb-spike).

Nine years after *Iguanodon* was first described and named, in May 1834, quarrymen at work in a limestone quarry near

81

Maidstone in Kent, observed some brown objects which they mistook for fossil wood. The quarry was owned by a Mr Bensted, who had an interest in fossils and recognised that they were pieces of large bones; he then directed that every piece should be collected. After much painstaking work, he obtained a considerable part of the disassociated skeleton of a young *Iguanodon*. Mantell was understandably excited by such a specimen, much more complete than all remains of that reptile hitherto found. Mr Bensted was willing to sell the fossil for £25, a sum that Mantell could not afford to pay. Finally, a group of Mantell's friends purchased the slab which contained the *Iguanodon* and presented it to Mantell (Colbert, 1968). Mantell's idea of the appearance of *Iguanodon* changed. In 1851 (p. 310) he had come to the conclusion 'that unlike the rest of its class, the Iguanodon had the body supported as in the mammalia, and the abdomen suspended higher from the ground than in any existing saurians'.

In the summer of 1832 Mantell again visited the quarry in Tilgate Forest which had yielded the first remains of *Iguanodon*. This time he was rewarded with the discovery of several bones indicating another kind of giant reptile, remarkable because of the extraordinary dermal bones and spines which covered its body, which he described as *Hylaeosaurus*.

Thus, at the beginning of the 1830s it was becoming clear that the world had once been inhabited, during the Jurassic and Cretaceous periods, by a number of strange extinct reptiles which had dwelled in the continents and the oceans, and with the pterodactyl had even become able to fly. The notion of an 'age of reptiles' became popularised by such authors as Mantell. As Cuvier expressed it in the last edition of his *Discours sur les révolutions de la surface du globe*, the class of reptiles had then 'taken its full development and exhibited various shapes and gigantic sizes'. This, of course, indicated a very different world from ours, one in which mammals had played an insignificant role in the economy of nature (although, as will be seen below, they were known to have coexisted with the giant reptiles). It also clearly implied that all those bizarre reptiles had become extinct, to be replaced by mammals. This was difficult to accept for those who still believed in the literal truth of Genesis, and the strangest explanations were put forward to get rid of this difficulty. Thus, in 1835 the Reverend William Kirby, an expert on entomology, published the seventh volume of the famous *Bridgewater Essays*, a series of eight books commissioned in the will of the Earl of

Bridgewater to demonstrate the 'Power, Wisdom and Goodness of God in the Works of Nature' (Bowler, 1984). Kirby's aim was to illustrate those divine virtues 'as manifested in the creation of animals', and he adhered as much as possible to a strict interpretation of Scripture. To the question whether any of the animals with which God had peopled the earth 'preparatory to the creation of man' were now extinct, he answered in the negative. This was unlikely, because Genesis clearly indicated that Noah had saved a pair of every living thing. The idea of an 'age of reptiles', as expressed by Mantell, was unacceptable to the Reverend Kirby (1835, p. 36), because it could not 'be reconciled with the account of the creation of animals as given in the first chapter of Genesis'. Fossils apparently did indicate that large reptiles had existed, but that they had characterised a definite period of earth history and had then become extinct could not be envisaged (Kirby, 1835, p. 39):

> who can think that a Being of unbounded power, wisdom and goodness should create a world merely for the habitation of a race of monsters, without a single rational being in it to glorify and serve him. The supposition that these animals were a separate creation, independent of man, and occupying his eminent station and throne upon our globe long before he was brought into existence, interrupts the harmony between the different members of the animal kingdom, and dislocates the beautiful and entire system, recorded with so much sublimity and majestic brevity in the first chapter of Genesis.

It was much more likely that giant reptiles had not become extinct at all. After all, there still were many unexplored regions on earth, where unknown animals might be lurking. However, as this old argument was becoming increasingly difficult to use, Kirby (1835, p. 20) proposed an additional and startling hypothesis: 'besides the unexplored parts of the surface of the earth and of the bed of the ocean, are we sure that there is no receptacle for animal life in its womb?' What he proposed was nothing less than the survival of the supposedly extinct giant reptiles in vast cavities inside the earth. 'Dragons' were traditionally associated with the 'abyss', and the living *Proteus anguinus* was an instance of a 'perfectly subterranean' reptile. Kirby was full of confidence when he wrote (1835, p. 36): 'I trust that my hypothesis of a subterranean metropolis for the Saurian,

83

and perhaps other reptiles, will not be deemed so improbable as it may at the first blush appear'.

Needless to say, his hypotheses met with little success. Mantell called them 'the most strange conceits', and remarked that 'as Dr Buckland's Essay follows that of Mr Kirby, the reader has the bane and the antidote both before him' (Mantell, 1838, p. 443). William Buckland had written the eighth Bridgewater Essay, on *Geology and Mineralogy Considered with Reference to Natural Theology* (1836), and the views he expressed there of course had little in common with those of the Reverend Kirby.

What the inhabitants of the world during the age of reptiles had been like was not perfectly clear, especially as far as the terrestrial forms were concerned. More or less complete skeletons of ichthyosaurs and plesiosaurs had been found, but *Megalosaurus, Iguanodon* and the like were represented by rather fragmentary remains. Reconstructions based on these were bound to be hypothetical, as witnessed by Cuvier's conception of *Megalosaurus* as a very large aquatic reptile, or Mantell's early reconstructions of *Iguanodon* as an iguana-like animal. As more complete specimens came to light, the ideas of palaeontologists about those giant reptiles gradually evolved. As mentioned above, the Maidstone find led Mantell to reconstruct *Iguanodon* as a more massive reptile, which was able to hold its body well off the ground on erect limbs. This trend culminated in Owen's definition of the Dinosauria in 1841.

Richard Owen (1804–92) was the leading British anatomist of the middle part of the nineteenth century. He had become assistant conservator of the Hunterian Museum (which belonged to the Royal College of Surgeons) in 1827, and this had provided him with an unequalled opportunity to study comparative anatomy. A visit to Cuvier in 1831 had further encouraged him to study fossil vertebrates, and he had soon become the foremost figure of British vertebrate palaeontology. At the 1841 meeting of the British Association for the Advancement of Science, Owen gave a review of British fossil reptiles in which he came to the conclusion that *Megalosaurus, Iguanodon* and *Hylaeosaurus* could not be referred to any group of previously known reptiles, and should be placed in a group of their own, for which he coined a name which was to prove very successful indeed (Owen, 1842, p. 103):

The combination of such characters, some, as the sacral ones, altogether peculiar among reptiles, others borrowed, as it were, from groups now distinct from each other, and all manifested by creatures far surpassing in size the largest of existing reptiles, will, it is presumed, be deemed sufficient ground for establishing a distinct tribe or suborder of Saurian Reptiles, for which I would propose the name of *Dinosauria*.

Owen's 'terrible lizards' were thus a group of very large extinct reptiles with peculiar anatomical features, which in some respects made them morphologically intermediate between various groups of otherwise distinct animals. The most important point was that they were now no longer merely giant lizards, but were envisaged as enormous reptiles with many mammalian features. In 1854 Owen's conception of what dinosaurs had looked like materialised in the famous Crystal Palace reconstructions. These were erected in the new grounds of the Crystal Palace, which had been the main attraction of the 1851 Exhibition in Hyde Park and had been rebuilt in Sydenham Park, in a suburb of south London. Owen was responsible for the scientific supervision of the project, while its material realisation was the work of the sculptor Waterhouse Hawkins. Among the life-size concrete models of prehistoric animals were not only the 'classical' extinct monsters of the nineteenth century, ichthyosaurs and plesiosaurs, but also Owen's Dinosauria: *Iguanodon, Megalosaurus* and *Hylaeosaurus*. They all shared a common general appearance: they appeared as very large, rather heavy-limbed, massively built quadrupeds. *Megalosaurus* was shown with a crocodilian-like head which befitted its carnivorous diet, while *Iguanodon* was given beak-like jaws surmounted by a horn, but these different heads were set on similar, somewhat rhinoceros-like bodies. In Owen's view, the dinosaurs were thus some kind of reptilian equivalent of the large mammals of today. This clearly made them superior to today's reptiles — in obvious contradiction to the interpretations of the fossil record which sought to establish a clear linear progression of the living world through geological time. As pointed out by Desmond (1982), Owen's reconstruction of the dinosaurs was probably part of his campaign against his scientific adversary, the Lamarckian Robert Grant.

The idea of a progression of life indicated by the fossil record (Bowler, 1976) became popular in some scientific circles as discoveries of older and more abundant fossils multiplied during the first half of the nineteenth century. Fossil vertebrates were of prime importance in the debate about progression, because they could be arranged in a hierarchical order more easily than the invertebrates, with the 'lowly' fishes at the bottom and the mammals at the top, amphibians and reptiles occupying inter-mediate positions. This amounted more or less to a chronological transposition of the 'great chain of being' popular among eighteenth-century naturalists, and it did not necessarily imply the idea of an evolution from the bottom to the top of the ladder, although it could obviously be used to support such a view. Such a progression could also be linked to changes in the physical conditions on earth. Buffon had already suggested such a connection between the supposed cooling of the globe and the successive appearance of various life forms, and similar, more elaborate views linking biological events to changes in earth history became popular in the first part of the nineteenth century. The still-limited number of fossil vertebrates known to Cuvier already suggested some kind of progression through geological time. In the *Discours sur les révolutions de la surface du globe*, he had pointed out that oviparous quadrupeds appeared much earlier than viviparous ones. Mammals were latecomers in the history of life, and among them there was a succession, the species still living today appearing only in the most recent deposits.

However, a considerable branch of vertebrate palaeontology had been left virtually untouched by Cuvier: although he had had plans to study fossil fishes, he never had time to prepare the great work which would have complemented his researches on fossil quadrupeds. To achieve what Cuvier had not been able to do was to be the greatest contribution of Louis Agassiz (1807–73) to vertebrate palaeontology. Agassiz was born in Switzerland and studied at the universities of Zürich and Heidelberg before moving to Munich in 1827. He first became interested in living fishes, and in 1829 published a description of the fishes collected by an expedition to Brazil. Having thus acquired a good knowledge of the anatomy of recent fishes, he turned to the study of the fossil forms. Encouraged in this by Cuvier, he visited natural history collections in various European countries and studied a very large number of fossil fish specimens. With the help of Alexander von Humboldt, he obtained a position at the Academy of Neuchâtel, in Switzerland, which he kept until his departure for the United States (where he eventually became a professor at Harvard) in 1846. His research resulted in the publication, between 1833 and 1844, of his *Recherches sur les poissons fossiles*, a five-volume work which for the first time presented a comprehensive review of fossil fishes. This was followed in 1844 by a monograph on the fossil fishes from the Old Red Sandstone, in which he described what were then the oldest known vertebrates.

Agassiz's work on fossil fishes (Gaudant, 1980) resulted in a classification of fishes based on scale morphology; however, the four orders (placoids, ganoids, ctenoids and cycloids) which he distinguished were soon found to be fairly artificial groupings, and his classification met with little success. He also attempted to use fossil fishes for purposes of palaeoclimatic reconstruction; the Eocene fish fauna of France and England, for instance, contained a large proportion of species belonging to genera found today in tropical regions. Most important, he found a clear link between the stratigraphic provenance of fossil fishes and their organisation and systematic position (Agassiz, 1833, p. VIII):

Fossil fishes differ according to the great geological formations in which they are found, and in each of them they have a particular kind of organisation which enables us to distinguish them. The older the rocks in which they are found, the more different they are from the fishes of the present world.

There was therefore some kind of development of fishes through geological time. This idea of progression could not be pushed too far, however. As a loyal disciple of Cuvier's principles, Agassiz could not admit that all groups of animals, invertebrates as well as vertebrates, could be included in a single chain, whether anatomically or chronologically. He thought that there was no evidence that vertebrates had appeared later than invertebrates. There was evidence of a 'regular organic development' among the vertebrates, but the invertebrates could not be included in the same process. Fossil and recent fishes could be arranged in a diagram which showed this development through geological time. In 1844 Agassiz presented his general conclusions in the form of such a 'genealogy of the fish class', showing the appearance, expansion and disappearance of the various groups on a geological time-scale. This diagram looks remarkably like more recent graphic depictions of the evolution of zoological or botanical groups — but Agassiz clearly stated that the various groups of fishes 'were not descended from each other by way of direct procreation or successive transformation' (1844, vol. 1, p. 171). There was no continuous descent from one type to another, but a 'reiterated manifestation of a predetermined order of things, tending toward a precise aim and methodically realised through time' (Agassiz, 1844, vol. 1, p. 171). In other words, living beings were the direct expression of the Creator's will, and regularities in their appearance and development testified to a well-ordained plan. The existence of a personal God was clearly manifested in the living world, and the final goal of the chronological development revealed by the fossil record was the creation of man. The idea of the transformation of species was a gratuitous hypothesis which opposed 'all sound physiological notions'. Agassiz never changed his mind on this point, and at the end of his career he was to become one of the leading, if unsuccessful, opponents to Darwinian evolution. This position prevented him from making full use of some of his discoveries. He had noticed, for instance, that there was some sort of parallelism between embryonal development and the successions of fossil forms through geological time. For him, however, this only showed that successive creations had occurred following a definite pattern, which was also recognisable in individual development. This parallelism again led him to the conviction that a divine creative will was behind all vital phenomena.

The idea of a more or less regular progression of fossil vertebrates through geological time was soon shaken by palaeontological discoveries. An early find of great importance was that of the famous Stonesfield mammals (Desmond, 1984). Around 1812 W.J. Broderip, then an undergraduate at Oxford, found two tiny jaws in the Stonesfield slate. He sold one to his tutor William Buckland, and mislaid the other, which was not described until 1827. Buckland and Broderip recognised that the jaws belonged to mammals. The presence of mammals in the Stonesfield slate ran counter to all expectations. The succession of vertebrate forms through geological time established by Cuvier and his immediate followers indicated that mammals had appeared much later, at the time of the deposition of the Tertiary strata. Some naturalists had even supposed that mammals would not have been able to breathe the Mesozoic atmosphere. That mammals had lived during the 'age of reptiles' was, to say the least, a disturbing fact, which had to be accounted for in some way.

One way out of the problem was to consider the Stonesfield mammals (widely held to be opossums) as a minor exception to a general rule. This was the position chosen by Cuvier, who had briefly seen one of the jaws during a visit to Oxford in 1818. In late editions of the *Discours* he mentioned that the jaws of an animal belonging to the didelphid marsupials seemed to have been found in the Oolite near Oxford. If it could be demonstrated that the fossils did come from Jurassic strata, this marsupial would be the oldest known mammal species, but such an isolated find would merely be a rather unimportant exception.

To other geologists and palaeontologists this was not a very satisfactory solution to what was perceived as a serious problem. The French geologist Constant Prévost tried in 1824 to reinterpret the Stonesfield slate as a post-Oolitic deposit: if the embarrassing marsupials could be shown to be contemporaneous, or nearly so, with the other fossil mammals described by Cuvier, the problem simply ceased to exist. This reinterpretation, however, was unacceptable to British geologists. If the age of the Stonesfield slate was indisputable, the only remaining solution was to deny the mammalian nature of the fossils. One of the first to do so was Robert Grant, who according to Desmond (1984) could not fit Mesozoic mammals into his materialistic Lamarckian scheme of regular ascent of living beings through geological time. In 1834 Grant stated that the attribution of the Stonesfield jaws to mammals was erroneous, an opinion which was accepted by Agassiz. These doubts about the zoological position of the fossils found an echo in France, where the anatomist Blainville, a noted anti-Cuvierian, decided that they actually belonged to reptiles. Although Blainville had not seen the original specimens when Buckland had brought them to France in 1838, he nevertheless stated that they had nothing to do with the marsupial *Didelphis*, and did not belong to insectivores either. If they actually were mammal jaws, their closest relatives would be seals. However, it was much more likely, according to Blainville (1838a,b), that they had to be attributed to saurians. As evidence for this, he compared the Stonesfield jaws, for which he proposed the generic name *Amphitherium*, with the remains of a mysterious fossil creature from Alabama which had just been described by the American palaeontologist Harlan under the name of *Basilosaurus*. This animal was thought to be a large reptile, but its teeth had complex roots, like those of the Stonesfield jaws. Blainville's conclusion was that the latter obviously did not prove that mammals had been in existence before the Tertiary.

There was considerable disagreement among French scientists concerning the systematic position of the Stonesfield animals, and a controversy developed at the Academy of Sciences. The ichthyologist Valenciennes, who had been Cuvier's collaborator, had seen the original specimens and had made casts of them. He thought (Valenciennes, 1838) that they belonged to mammals allied to *Didelphis*, and proposed the generic name *Thylacotherium* for them. The herpetologist Duméril (1838) was also convinced that they belonged to mammals because of their articular condyle; he also remarked that the vertebrae of *Basilosaurus* were very similar to those of a cetacean. Etienne Geoffroy Saint-Hilaire's opinion (1838) was an attempt to solve the problem in an unexpected way: he accepted that the Stonesfield jaws belonged to marsupials, but he claimed that marsupials could not be considered as mammals! In this way regular progressionism was saved: there were indeed marsupials in Secondary rocks, together with giant reptiles, but true mammals did not appear until the Tertiary.

In Britain, Buckland found an important ally in the person of Richard Owen, who shared his dislike of Grant's radical views. In 1838 Owen presented his conclusions to the Geological Society of London: the Stonesfield jaws were definitely mammalian and could be considered as belonging to an insectivorous marsupial (1838a). This failed to convince supporters of the 'saurian' hypothesis, who still invoked the resemblances with *Basilosaurus*. In 1839 Harlan arrived in Britain with some *Basilosaurus* bones which were studied by Owen and Buckland. They confirmed Duméril's suspicion that the animal was actually a kind of whale, and Harlan easily accepted that he had been mistaken. In 1841 he published a recantation of his former views, and Owen thought it necessary to rename the animal *Zeuglodon*. The last argument in favour of the reptilian nature of the Stonesfield jaws thus disappeared, although Grant never accepted that they belonged to marsupials (Desmond, 1984). The existence of small mammals during the 'age of reptiles' was demonstrated, and the simpler versions of progressionism which could not accommodate mammals prior to the Tertiary, were discredited.

During the first half of the nineteenth century, opposition to progressionism was bolstered by the emergence of uniform-itarian geology, which was based on the assumption that the causes which had acted in the past were not different from those in operation today. Such 'actualist' concepts had supporters on the Continent, among them the German K.E. von Hoff (Wendt, 1971) and the Frenchman Constant Prévost, but the most influential proponent of uniformitarian ideas was undoubtedly Charles Lyell (1797–1875). Before he became a reluctant supporter of evolution in the 1860s, Lyell's conception of a 'steady-state' earth had led him to a strongly anti-progressionist position. If physical conditions on earth had always been similar to those of the present day, it was logical to assume that living populations had also been similar to those of today. Lyell, however, accepted that many species had become extinct, but he did not believe that the earth had once been inhabited only by simple forms of life which had been succeeded by more complex ones. In his steady-state view, as Ruse (1979) has called it, earth history was thought to be cyclical, and living beings were also involved in these cycles. Species were created and became extinct as part of a regular process. In the first volume of his *Principles of Geology* (1830), Lyell even went so far in this direction as to suggest (p. 123) that, after a complete cycle had been gone through, a time might come when 'the huge iguanodon might reappear in the woods, and the ichthyosaur in the sea, while the pterodactyl might flit again through umbrageous groves of tree-ferns'. This extreme form of anti-progressionism prompted a reply from Henry De la Beche in the form of his famous lithographic caricature entitled 'Awful Changes', in which a professorial ichthyosaur (Lyell) demonstrates to ichthyosaur students that the human skull provides clear evidence of a low order of organisation (McCartney, 1977).

Whereas the Lamarckians, when trying to recognise a simple progression in the fossil record, were led to dismiss evidence contrary to this view, Lyell and his followers were faced with the opposite problem, viz. the lack of evidence for 'higher' organisms in the older geological formations. Here again, fossil vertebrates could provide choice material for a demonstration of

91

the non-progressive nature of the fossil record: all remains of higher vertebrates found (or supposedly found) in rocks previously thought to contain only 'lowly' animals were presented as evidence in favour of the steady-state view of earth history. Thus, Lyell of course used the Stonesfield mammals as an especially clear vindication of his ideas. Similar instances from older rocks were even more welcome, and when an advanced reptile was reported from the Old Red Sandstone of Scotland, Lyell grew understandably enthusiastic about the new find (Bowler, 1976; Benton, 1982). The specimen, an incomplete skeleton with part of a skull, had been found in 1851 in a sandstone quarry near Elgin. The vertebrate-bearing sandstones in the Elgin area were generally thought to be Devonian in age. Reptile plates from these rocks had been misinterpreted by Agassiz as those of a ganoid fish typical of the Old Red Sandstone, and it was not until about 1860 that the discovery of several reptiles (thecodontians and rhynchosaurs) in the Elgin sandstones finally convinced geologists (including Lyell) that they actually belonged to the Triassic. In 1851 the fossil found near Elgin could still be interpreted as proof that reptiles had been in existence as early as the Devonian. The specimen had been acquired by a Patrick Duff of Elgin, who sent drawings of it to Owen. However, Duff's brother-in-law, Captain Lambart Brickenden, was a friend of Gideon Mantell, and endeavoured to obtain the specimen for him. Lyell heard about the discovery from Mantell and was much interested in it, to the extent that he added a footnote on it to a new edition of his *Manual of Elementary Geology*. Before Mantell and Brickenden could publish their description of the animal as an amphibian, which they named *Telerpeton elginense*, Owen was able to publish his own description of the fossil, which he identified as a lizard and called *Leptopleuron lacertinum*. Mantell, who had already violently disagreed with Owen on other points, reacted strongly to this, and regarded Owen's attitude as unethical, although it does seem that Owen had been asked by Duff to describe the specimen well before Mantell and Lyell heard about it (Benton, 1982). In any case, Lyell's insistence that the specimen should be described by Mantell was probably due to his desire to oppose Owen's version of progressionism. In Owen's conception of vertebrates as variations on a basic archetypal 'theme', progression through geological time was not the simple unilinear development of the Lamarckians but a much more flexible diversification process, which could

Figura Sceleti prope Qvedlinburgum effossi.

Fig. 1: The influence of the unicorn myth on early reconstructions of vertebrate fossils: Otto von Guericke's drawing of a fossil skeleton, based on mammoth bones from Quedlinburg (after Leibniz, 1749).

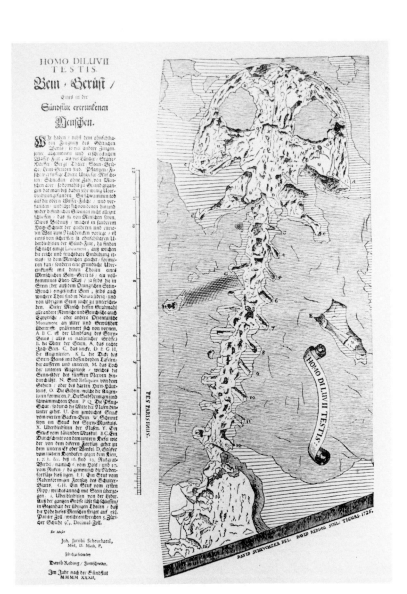

Fig. 2: Scheuchzer's *Homo diluvii testis*: a giant salamander from the Miocene of Oeningen.

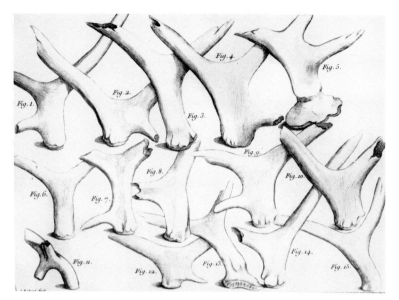

Fig 3: Exotic animals in the ground of western European countries: reindeer antlers from Etampes, figured by Guettard (1768).

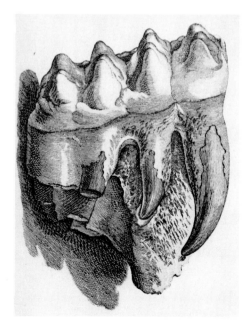

Fig 4: The "American Incognitum": a mastodon tooth from North America, figured by Buffon (1778).

Fig. 5: A famous episode of eighteenth-century palaeontology: the discovery of the "Great Animal of Maastricht" in the underground quarries of Saint Peter's Mountain (Faujas de Saint-Fond, 1799).

Fig. 6: A portrait of Georges Cuvier, from the last, posthumous, edition of his
Recherches sur les ossemens fossiles (1834–36).

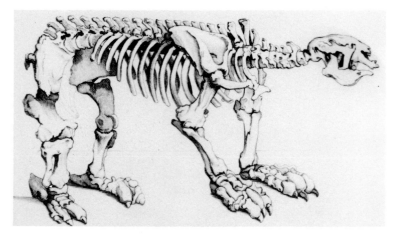

Fig. 7: The "animal from Paraguay", actually a giant ground sloth from the Pleistocene of Argentina, which was mounted in Madrid by Bru, and described by Cuvier as *Megatherium*.

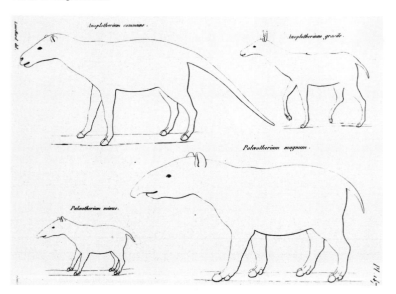

Fig. 8: The earliest scientific reconstructions of fossil vertebrates: Eocene mammals from the Montmartre gypsum, drawn by Laurillard under Cuvier's supervision (from Cuvier, 1834–36).

Fig. 9: The "age of reptiles" as envisioned by Thomas Hawkins (1840): a fight between an ichthyosaur and two plesiosaurs on the Jurassic seashore.

Fig. 10: Fossil reptiles in semi-popular books of the nineteenth century: a Romantic vision of an ichthyosaur skeleton surrounded by the "kobolds of German legends" (Pouchet, 1872).

Fig. 11: Owen's conception of quadrupedal dinosaurs: a fight between *Iguanodon* and *Megalosaurus*, as seen by the artist Riou in Figuier's *La Terre avant le Déluge* (1866).

Fig. 12: A progressionist view of the succession of living beings through geological time (Flammarion, 1886).

Fig. 13: Discovering the palaeontological riches of the American West: an exploring party in the *Mauvaises Terres* of Nebraska (from Leidy, 1853).

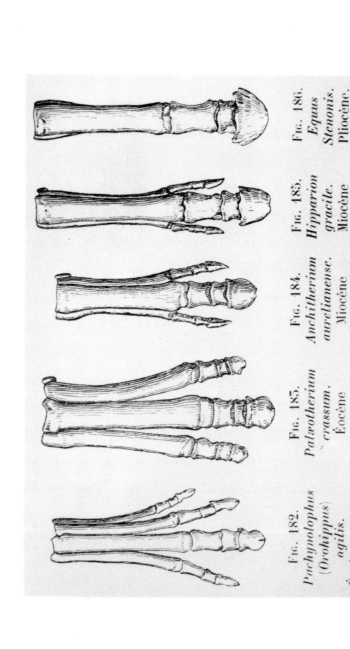

Fig. 182.
Pachynolophus
(Orohippus)
agilis.
Éocène moyen.

Fig. 183.
Palæotherium
crassum.
Éocène
supérieur.

Fig. 184.
Anchitherium
aurelianense.
Miocène
moyen.

Fig. 185.
Hipparion
gracile.
Miocène
supérieur.

Fig. 186.
Equus
Stenonis.
Pliocène.

Fig. 14: A late nineteenth-century "pseudo-phylogeny": stages in the evolution of the horse's foot, after Gaudry (1896).

Fig. 15: The bipedal dinosaurs: a "post-Bernissart" *Iguanodon* fights a *Megalosaurus* (Flammarion, 1886).

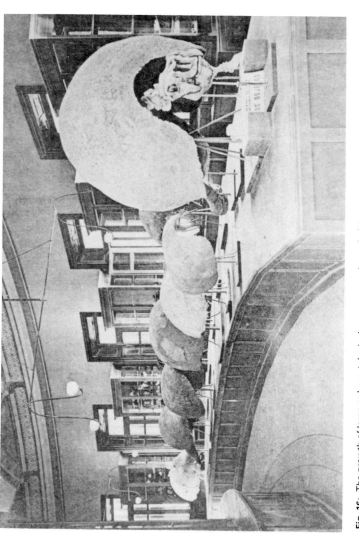

Fig. 16: The growth of large palaeontological museums at the end of the nineteenth century: the glyptodont collection in the La Plata Museum (from Moreno, 1891a).

M. COGESTALL
Aide-Naturaliste
Américain

M. HOLLAND
Directeur
du Musée Carnégie
à Pittsburg (Etats-Unis)

M. BOULE
Professeur au Muséum
d'Histoire Naturelle

864. - PARIS. - Jardin des Plantes - Galerie de Paléontologie
Montage du Squelette du Diplodocus, don de M. André Carnégie

Fig. 17: Carnegie's gift: a cast of *Diplodocus carnegiei*, from the Jurassic of Wyoming, being mounted in the Paris Museum under Holland's and Boule's supervision, in 1908 (from a postcard, courtesy Yannicke Dauphin).

Fig. 18: The acme of colonial vertebrate palaeontology: the German excavations of Jurassic dinosaurs at Tendaguru, East Africa (from Janensch, 1914).

explain many seeming 'irregularities' of the fossil record. Neither the Lamarckian view, nor Owen's more adaptable attitude, however, could easily be reconciled with Lyell's conceptions.

Despite Lyell's eagerness to find fossil evidence against progressionism, it proved difficult to discover actual remains of 'advanced' vertebrates in ancient rocks. Less tangible evidence had to be resorted to, and fossil footprints were found to be especially convenient in this regard, although admittedly they were not as demonstrative as bones. The most spectacular example of this was the controversy about the origin of the mysterious footprints which were described as *Chirotherium* (complete bibliography in Soergel, 1925). The story began in 1833, when F.K.L. Sickler, director of the *Gymnasium* at Hildburghausen, in Thuringia, was having a wall built in his garden. Unusual footprints were detected on a sandstone slab from the nearby Hessberg quarry. Sickler asked the workmen to look for more, and wrote a letter to Blumenbach to inform him of the discovery. The letter was eventually published in the *Neues Jahrbuch für Mineralogie* (1835) with remarks from the palaeontologist H.G. Bronn, who thought that the impressions had been left by some kind of monkey. The reason for this interpretation was the strangely hand-like appearance of the prints, with four forward-pointing 'fingers' and a lateral 'thumb' (the fact that the thumb was on the outer side of the 'hand' was apparently not thought to be important). As the impressions left by the front and hind feet had the same general shape, it seemed logical to attribute them to an animal with hand-like hind feet, and a monkey seemed a likely candidate. This was not the only possible interpretation, as shown by the discussion which followed in scientific journals, and found its way to the general public (in 1856 the German poet Eduard Mörike even alluded to the controversy in verse: see Martin, 1966).

Many eminent palaeontologists felt impelled to give their opinion on the origin of the footprints, which shows that they were thought to be of great potential importance. The Hessberg sandstones were known to be Triassic in age, and the existence of a monkey, or at least some kind of mammal, as long ago as the beginning of the Mesozoic was obviously a severe blow to progressionist conceptions. Alexander von Humboldt, in 1835, accepted that the mysterious animal was a mammal, but he was not sure what kind of mammal it was, and suggested a phalanger or a loris. A.F.A. Wiegmann (1835) was even more perplexed, as he compared the Hildburghausen footprints with those of the opossum, the bear, and even the crocodile. The famous German palaeontologist Kaup finally gave the mysterious creature a name in 1835; he thought it was probably 'a gigantic marsupial with thumbs on the fore and hind feet', and called it *Chirotherium Barthii* (after a certain Barth who had been instrumental in the discovery), with the proviso that if the creature turned out to be an 'amphibian' the name could be changed to *Chirosaurus*. Geoffroy Saint-Hilaire (1838) followed Kaup's diagnosis, which enabled him to push the appearance of his non-mammalian marsupials farther backward in time than the 'Oolitic' period in which the Stonesfield marsupials had lived. Kaup's interpretation was not accepted by F.S. Voigt, who rejected the idea of a marsupial and concluded that the footprints must have been left by a 'colossal monkey', a *Palaeopithecus*. This proved that mammals had been in existence when the early Triassic Buntsandstein had been formed, and led to even farther reaching conclusions (Voigt, 1835, p. 324):

> I believe that the human genus itself already existed in the earliest times (by which I mean the oldest neptunian period), but that it was hidden, possibly in some remote corner of the earth, like other animals.

After catastrophes had caused the emergence of larger continental areas, man could have spread to other regions. This conclusion was in complete opposition to all progressionist concepts, which considered that man, the 'summit' of creation, was the latest comer in a progressive succession. In 1836 Voigt returned to the *Chirotherium* question with a new, experimental approach: when an itinerant circus came to Weimar he had the opportunity of examining the soles of the feet of various exotic animals, including a bear and a mandrill (the ground unfortunately was too dry to allow footprints to be made). After this experiment Voigt was no longer so sure about his monkey hypothesis. He suggested that some thumbless prints may have been left by a bear, 'possibly the famous *Ursus spelaeus* itself'. Other footprints made him think, not entirely seriously, of a colossal frog. In any case, he rejected the possibility of a marsupial, and comparison with the mandrill revealed some differences. Voigt's conclusion was that in all likelihood this prehistoric animal no longer existed.

Other naturalists did not believe that the *Chirotherium* footprints had been left by a mammal, and chose interpretations less damaging to progressionist ideas. Thus Link, in 1835, thought that their author may have been an amphibian or a reptile, perhaps some kind of gigantic salamander. The amphibian hypothesis was to be developed by Richard Owen, who in 1841 in a study of the large fossil amphibians called *Labyrinthodon*, suggested that the name *Cheirotherium* (as he spelled it) should be considered as a synonym of *Labyrinthodon*. This interpretation became widely accepted, even by Lyell, who in his *Elementary Manual of Geology* (1855 edition) gave a reconstruction of *Chirotherium* as a large amphibian. The skeleton of the labyrinthodonts was still imperfectly known, and *Chirotherium* was given a rather frog-like appearance, with a very short tail and long limbs, which it supposedly used in a peculiar cross-legged fashion, to account for the peculiar position of the 'thumb'. This reconstruction of a labyrinthodontine *Chirotherium* was reproduced in many textbooks and popular works on palaeontology. It was not universally accepted, however. In 1858, for instance, when *Chirotherium* footprints were found in eastern France, the French geologist Daubrée studied the very fine impressions left by the surface of the skin and thought that they were more reminiscent of the footpads of mammals than of the scaly or smooth skin of reptiles or amphibians. This suggested to him that mammals may indeed have existed at the beginning of the Triassic. Nevertheless, the mammalian hypothesis was generally abandoned in the second half of the nineteenth century as the controversy about progressionism gradually died down; and the amphibian hypothesis fared little better. In 1867 W.C. Williamson described *Chirotherium* footprints from the Triassic of Cheshire, and his conclusion was: 'I do not believe that it belonged to a Batrachian animal. It is Saurian, if not Crocodilian, in every feature.' Although its precise systematic allocation was still in doubt, the reptilian nature of the footprint-maker was accepted by an increasing number of palaeontologists, until in 1925 the German palaeontologist W. Soergel could demonstrate convincingly that the likeliest candidates were pseudosuchian thecodonts, an opinion which is now generally accepted (Krebs, 1966).

In the debate about progressionism, the earliest known vertebrates were of obvious importance: according to the progressionists, they had to be simple primitive forms. Most of the evidence concerning early vertebrates came from the Devonian Old Red Sandstone of Scotland, where many dedicated collectors had been at work (Andrews, 1982). Agassiz, who had visited Scotland twice, had thus been able to study a large number of specimens. Although he was one of the foremost proponents of 'transcendental' progressionism (in which species were created in ascending order, the process culminating with the appearance of man), some of the Devonian fishes he worked on were not easy to fit into a simple scheme of progressive elevation within the vertebrates, or even within the fishes. Many of the Old Red Sandstone fishes and agnathans were highly ossified creatures, which could be interpreted as more 'advanced' than the later, less ossified fishes. Moreover, the Palaeozoic 'sauroid' fishes, as they were then called, showed suspect resemblances to amphibians or reptiles. These peculiarities of early fishes were used to dismiss the idea of progression within the whole class by Hugh Miller, a Scots stonemason who had become interested in geology, had collected many fossils and finally wrote several successful popular books on palaeontology and geology. In his *The Old Red Sandstone*, published in 1841, Miller argued that late Silurian and Devonian fishes were too highly organised to be considered as a link between invertebrates and later fishes in a continuous ascending progression. As pointed out by Bowler (1976), Miller, who had strong religious convictions, was worried that the progressionist view of the development of life could be used as evidence for a truly genealogical, i.e. evolutionary, link between organisms, which would have threatened the unique position of man in the living world. This fear was justified, for progression was indeed used in such a way by various evolutionists in the first half of the nineteenth century. Even though it had become very difficult to envisage a simple unilinear progression of the kind accepted by some Lamarckians, the increasing diversification of life through geological time revealed by palaeontological discoveries posed a serious problem to those who still hoped to explain the origin of every species through a special act of divine creation. It had become completely unrealistic to postulate that all living beings had been created at the same time, and the remaining alternative was to suppose that successive creations had occurred. If species were not able to change and transform into other species, a mechanism also had to be found to explain the disappearance of the enormous number of extinct species revealed by palaeontology.

This led to theories of successive creations and catastrophic extinctions, as exemplified by the conceptions of Alcide d'Orbigny (1802–57), a French invertebrate palaeontologist. In his *Cours élémentaire de paléontologie et de géologie strati-graphique* (1849–51), d'Orbigny clearly rejected progressionism. He could not find any regularity in the development of the various animal groups: at a given period, some were expanding while others were on the decline, and more advanced forms had often been replaced by less 'perfect' ones. *Parallel* changes could be recognised, but progressionism could not be supported by the fossil record. What the palaeontologist observed was (d'Orbigny, 1851, p. 251): 'the succession, in a constant chronological order, in all parts of the world, of distinct faunas for each stage, which have replaced each other, from the first animalisation of the globe to the present'.

The creative force responsible for these appearances was beyond human understanding, 'there are limits the human mind cannot cross'. A first creation had taken place with the Silurian and had been destroyed by geological processes, after which a second creation had occurred at the beginning of the Devonian, and 'successively twenty-seven times distinct creations have repeopled the whole earth with plants and animals, following each geological perturbation which had destroyed everything in living nature'. This, admittedly, was incomprehensible and shrouded in superhuman mystery, but it was a fact. D'Orbigny had no doubt that there was no transition between the fauna of one geological stage and that of the next; global extinction was the rule. The cause of these extinctions could not be sought in mere climatic change; one had to envisage much more violent geological perturbations, which had dislocated the earth's crust on a worldwide scale and caused widespread changes in the distribution of land and sea. This insistence on the role of tectonic disturbances was in some respects a continuation of Cuvier's ideas (although Cuvier did not believe in multiple creations), and was in agreement with the views of Elie de Beaumont, one of the leading French geologists of the time, who had elaborated a theory of successive worldwide tectonic phases.

D'Orbigny's rather extreme views were representative of a fairly successful current of palaeontological thought, which at one time seriously threatened early evolutionary interpretations.

In 1843 the Belgian geologist J.J. d'Omalius d'Halloy defended a rather Lamarckian theory of progressive change by way of reproduction (under the influence of environmental modifications), but had to admit in a footnote (p. 726) that the doctrine of successive creations had been gaining ground since 1831 (when he had first published his evolutionary ideas), while that of gradual change had grown less convincing. He still had doubts about the abrupt nature of the separation between fossil faunas, however, and was not yet ready to give up his evolutionary views completely.

D'Omalius d'Halloy's attitude may have been characteristic of the perplexity of many geologists and palaeontologists in the 1840s and 1850s. The unilinear progression of living beings which some Lamarckians had hoped to discover in the fossil record had not been confirmed, but the increasing diversification of the organic world through geological time with, especially among the vertebrates, a successive appearance of the major groups, seemed to many naturalists to have become an indisputable fact. This was expressed in clear terms by H.G. Bronn in his essay on the distribution of fossil organised bodies in the sedimentary rocks (1861) which he submitted to the French Academy of Sciences as a response to a prize offer of 1850. Bronn suggested that the development of life had been conditioned by two fundamental laws. One was that the appearance of given types of organisms was directly linked to environmental conditions: animals and plants appeared only when the appropriate exterior conditions had been realised. The second law was that of progressive development, which did not apply to the organic world as a whole, but was clearly demonstrated within each group of living beings. Bronn insisted that the successive changes in the living populations were the result of the extinction of ancient species and the creation of new ones, and were not due to the transformation of one species into another. The absence of intermediate fossil forms was cited as evidence in favour of the fixity of species. Species had become extinct when they ceased to be adapted to their environment, and had been replaced by new, better-adapted ones. Bronn was convinced that 'each species is the effect of a new act of creation (whatever the idea one has about it), and wherever new species are born, creation is still active' (Bronn, 1861, p. 659). Although in both theories each species was the result of special creation, the major difference with d'Orbigny's views was that according to Bronn the creation of species was a more or less continuous phenomenon which ensured gradual faunal and floral replacement during a geological history from which universal catastrophes were banned. The cooling of the earth and the chemical evolution of its atmosphere were the determining factors which had conditioned organic change, and creative activity was not limited to brief periods after each 'revolution' as in d'Orbigny's hypothesis.

Bronn envisaged two ways in which species may have been created. Either each species had been specially created by a personal God who had decided their order and place of appearance from the beginning, or an unknown force had been responsible, under special laws, for the appearance and coordination of all animal and vegetal species. This force must have had close links with the inorganic forces which had conditioned the progressive development of the earth's surface, which explained the constant adaptation of living beings to their environment. Bronn favoured the latter interpretation: the idea of a Creator conducting the development of organic nature through a force placed within nature itself (just as the inorganic world was led by the force of attraction) was, to him, 'more sublime' than that of continual direct interference. The idea of God as a gardener cultivating his garden had little appeal for Bronn.

This conception of 'creation by law', rather than through miraculous intervention, was shared by various scientists of the time. Owen's archetypal anatomy also involved natural forces which directed the realisation of diverse morphological adaptations from the basic plan of the vertebral archetype (Ruse, 1979). What these creative forces were remained obscure, however. Evolutionary hypotheses had the merit of proposing less abstract mechanisms for the origin of species, whether by environmental influence, individual effort of the organism, or prolongation of embryonic growth (as advocated in Robert Chambers's anonymously published *Vestiges of Creation* of 1844). However, the early evolutionary theories were unable to gain widespread support in the scientific community, not only because they went against religious and philosophical beliefs, but probably because they also failed to provide fully convincing mechanisms of species formation. The superiority of Darwin's hypothesis of natural selection largely lay in this field.

6

Vertebrate Palaeontology and Darwinism

During the first half of the nineteenth century, palaeontology, and especially vertebrate palaeontology, greatly contributed to the formation of an intellectual climate in which evolutionary ideas could be accepted. The fact that the organic world of today was only the last stage in an extremely long succession of organisms — many of which had become completely extinct, while others showed strange resemblances with living forms — had made the idea of a single act of creation at the beginning of earth history untenable. Without the discoveries of palaeontology, the question of the origin of species would not have become as pressing as it was in the 1850s. However, when evolutionary conceptions finally gained the upper hand with Darwin's natural selection hypothesis, palaeontology's contribution to this victory was relatively limited. This was due to the very nature of palaeontological evidence: fossils could be used to reconstruct an evolutionary history, but they gave few indications about the causes and mechanisms of evolutionary change. Moreover, in the 1850s the gaps in the fossil record were still considerable, which made widely divergent interpretations possible.

Nevertheless, fossil vertebrates did play a part in the development of Charles Darwin's ideas on evolution. Darwin's interest in vertebrate palaeontology was manifested by his own discoveries in South America during the voyage of HMS *Beagle* between 1831 and 1836 (Darwin, 1845; page numbers of quotations from the 1959 edition). In August 1833 the *Beagle* arrived in Bahia Blanca, on the coast of Argentina, and Darwin was left behind 'with captain Fitz Roy's consent' to collect fossils from the Pampaean Formation, a Pleistocene deposit of clays and marls which extends over a vast area of the Argentinian pampas. In a place called Punta Alta, Darwin found abundant remains of gigantic land mammals, which were eventually studied by Richard Owen. In his account of the voyage of the *Beagle*, Darwin mentioned ground sloths (*Megatherium, Megalonyx, Scelidotherium, Mylodon Darwini*, named in his honour by Owen, and another gigantic edentate), a large animal with an armadillo-like armour, an extinct kind of horse, a tooth of *Macrauchenia* and remains of *Toxodon*. Thirty miles (48km) from Punta Alta, more bones were discovered, among them teeth of a large capybara-like rodent, and part of the head of the rodent *Ctenomys*, a form close to the living *tuco tuco*. Darwin noted that the Punta Alta vertebrate remains were found in stratified gravel and mud, together with 23 species of shells, of which 13 were recent and 4 closely related to recent forms, while the others were either extinct or unknown. Using Lyell's stratigraphic methods, by which Tertiary deposits had been divided into several epochs on the basis of the percentage of recent invertebrates they contained, Darwin came to the conclusion that the Punta Alta deposit belonged to 'a very late tertiary period'. This confirmed Lyell's idea that 'the longevity of the species in the mammalia is upon the whole inferior to that of the testacea'. However, at the time he wrote the *Journal of Researches*, Darwin was not soley influenced by Lyell; in the field of vertebrate palaeontology, he relied much on Owen's anatomical knowledge. His description of the life-habits of the giant sloths he had found was based on Owen's interpretations and reconstructions, which, among other results, showed that these animals, unlike their small, present-day cousins, had been too large to climb trees, and must have pulled the branches down with their arms to feed on them. This led Darwin to interesting palaeoecological considerations. He at first thought that the huge herbivores found in the Pampaean Formation needed a much more luxuriant vegetation than that of the present pampas. He soon changed his mind, and came to the conclusion that the plains scattered with thorny trees found a little farther north could have supported large populations of herbivorous mammals. He drew conclusions from his own observations at the Cape of Good Hope, and from those of travellers who had visited the interior of southern Africa: obviously, plains covered with 'a poor and scanty vegetation' provided enough food for large numbers of big mammals such as elephants, rhinoceroses, giraffes and buffaloes. Conversely, the lush vegetation of the tropical parts of South America did not support many large mammals. This brought Darwin to the old problem of the Siberian frozen fauna: since large herbivores were not necessarily indicative of a lush tropical vegetation, there was no reason to assume that the large fossil mammals from Siberia could not have lived in the places where their remains were found. Therefore, there was no need to postulate sudden climatic catastrophes to account for the occurrence of elephants and rhinoceroses in the frozen ground of Siberia.

Not only were fossil mammals no proof of past catastrophes, they also suggested some unexpected relations between living forms belonging to different zoological groups. Drawing on Owen's description, Darwin remarked that the *Toxodon* (now known to be an endemic notoungulate) showed close affinities with the 'gnawers' (rodents), whereas some anatomical features also allied it with the 'pachyderms', and the position of the eyes indicated relations with the dugong and the manatee. The comparisons were rather fanciful, but they did lead Darwin to far-reaching conclusions: 'How wonderfully are the different Orders, at the present time so well separated, blended together in different points of the structure of the Toxodon!' This kind of observation would later be used as evidence for evolutionary divergence from a common ancestral stock.

After excavating fossil vertebrates at Bahia Blanca, Darwin rode to Buenos Aires, and from there he set out on an excursion to Santa Fé, on the Parana. On 1 October 1833 he arrived at a river, the Rio Tercero or Saladillo, and spent the greater part of the day there searching for fossil bones. Besides a *Toxodon* tooth and many scattered bones, he found 'two immense skeletons' projecting from a cliff. They were too decayed to be collected, but a tooth fragment allowed him to identify them as belonging to mastodons. As mentioned in Chapter 1, the local people interpreted them as remains of gigantic burrowing animals.

In Santa Fé, Darwin was delayed for five days, which gave him the opportunity to study the geology of the surroundings. He found evidence of gradual elevation within the recent period, and discovered 'the osseous armour of a gigantic armadillo-like animal, the inside of which, when the earth was removed, was like a great cauldron'. Besides this glyptodon, he also collected teeth of *Toxodon* and mastodons, as well as a horse tooth, which greatly interested him for biogeographical reasons. The discovery of fossil horses in America was of course unexpected, as this animal no longer existed there when the first European colonists arrived. Darwin reflected on the former resemblances between the South American and North American faunas, which in his opinion were greater than the present ones. This, to him, was a remarkable instance 'where we can almost mark the period and manner of the splitting up of one great region into two well characterized zoological provinces'. He thought that this relatively recent splitting-up was the consequence of the elevation of the Mexican platform, or more probably, of the submergence of land in the West Indies. He went on to examine the former close resemblances of the North American mammal fauna with that of Europe and Asia, and came to the conclusion that the north-western side of North America, in the region of Behring's Straits, was 'the former point of communication between the Old and so-called New World', an opinion which had been anticipated by Buffon and which has been confirmed by subsequent work. From North America several groups of land mammals had entered South America to mingle with the endemic fauna, after which a wave of extinction had occurred. In the 1840s Darwin thus had a fairly exact general conception of what would later be called the 'Great American Interchange'.

Other observations in the region of Santa Fé led him to anti-catastrophist reflections: a few years before, a great drought had affected the country and millions of animals had died. Exhausted cattle herds which had rushed into the Parana had been drowned there, and thousands of carcasses must have been deposited in the estuary and buried there by subsequent floods. Such an event was part of the 'common order of things', but its results (viz. 'an enormous collection of bones, of all kinds of animals and of all ages, thus embedded in one thick earthy mass') could easily be interpreted by a later (and presumably non-Lyellian) geologist as the consequence of a vast flood sweeping over the surface of the land.

In November 1833, when he was travelling to Montevideo, Darwin again had the opportunity to obtain fossil mammal remains. Having heard of giant bones at a nearby farm, he rode there and was able to purchase what had been a fine *Toxodon* skull (unfortunately, boys had 'knocked out some of the teeth with stones, and then set up the head as a mark to throw at'). He also found glyptodon remains and a partial *Mylodon* skull. It turned out that the remains of large extinct mammals were extraordinarily abundant in the pampas. Many local geographical features were named after them, with names such as 'hill of the giant' or 'stream of the animal', and there were legends about 'the marvellous property of certain rivers, which had the power to change small bones into large'.

The *Beagle* then sailed southward along the coast of Patagonia, and there Darwin was again able to enlarge his collection of fossil mammals. At a place called Port St Julian, he discovered a partial skeleton of a 'remarkable quadruped', which was to be named *Macrauchenia patachonica*. The animal was as large as a camel and, following Owen, Darwin thought it belonged to the pachyderms, but also showed clear relations with the guanaco. Local stratigraphy showed that *Macrauchenia* had lived at a relatively late geological period.

When considering the general character and affinities of the extinct mammalian fauna which he had discovered in South America, Darwin came to an important conclusion: these extinct animals showed clear relationships with the living South American mammals. Among rodents, some extinct species even belonged to still-extant genera. The extinct edentates were related to the living ones, so characteristic of the South American fauna. Darwin went so far as to suggest that the *Macrauchenia* was distantly related to the guanaco, and the *Toxodon* to the capybara. His own views were confirmed by the discoveries of the Danish explorers Lund and Clausen in Brazilian caves. A similar resemblance could be found between fossil and living marsupials in Australia. Although he did not elaborate much on this point in the *Journal of Researches*, Darwin did indicate that these resemblances were of major importance for the question of the origin of species (p. 165):

This wonderful relationship in the same continent between the dead and the living, will, I do not doubt, hereafter throw more light on the appearance of organic beings on our earth, and their disappearance from it, than any other class of facts.

The question which arose next was that of extinction: what had caused the disappearance of all the strange, large animals which had formerly inhabited South America? At first sight, a great catastrophe, able to 'shake the entire framework of the world', would seem to be needed to exterminate so many genera and species. However, the geology of La Plata and Patagonia revealed only slow and gradual changes. Climatic change was an unlikely cause for extinctions which had destroyed at the same time the inhabitants of tropical, temperate and arctic latitudes. Similarly, the action of man could explain the disappearance of some of the larger species, but was unlikely to have caused that of the small rodents. And it was difficult to imagine that the living species, much smaller than the extinct ones, had driven them to extinction by consuming their food. Extinction obviously remained a major problem: 'certainly, no fact in the long history of the world is so startling as the wide and repeated exterminations of its inhabitants' (Darwin, p. 166).

However, if one looked at the problem from another point of view, extinction could become less of a mystery. The conditions of existence of every species played an important role, but they were usually poorly known. Another important factor was that 'some check is constantly preventing the too rapid increase of every organised being left in a state of nature' (p. 166). Malthus's principles, applied to the animal world, showed that while the supply of food remained more or less constant, the propagation of every animal by way of reproduction was geometrical. Increase in numbers therefore had to be checked by some natural means — but the nature of the check was seldom known with any precision. Slight environmental differences could obviously exert a profound influence on the abundance of animal species. Processes of extinction certainly were similar, whether extinction was caused by man or by natural agents: in both instances a species first became rarer and rarer, and then disappeared altogether. A species became rare when its conditions of existence became unfavourable, and thus even slight environmental deterioration could lead to extinction. To these reflections brought about by the extinct mammals of South America, Darwin gave the following definitely anti-catastrophist conclusion (p. 168):

> To admit that species generally become rare before they become extinct — to feel no surprise at the comparative rarity of one species with another, and yet to call in some extraordinary agent and to marvel greatly when a species ceases to exist, appears to me much the same as to admit that sickness in the individual is the prelude to death — to feel no surprise at sickness — but when the sick man dies, to wonder, and to believe that he died through violence.

105

In the first edition of Darwin's *Origin of Species by Means of Natural Selection* (1859, page numbers of quotations from the 1968 edition), his observations on the extinct mammals of South America were used as evidence in favour of evolution. A whole chapter was devoted to the geological succession of organic beings. Darwin did not believe in a regular and fixed law of development which would have caused simultaneous changes in the living world. Species evolved at very variable rates, and this explained some apparent anomalies of the fossil record, such as the presence in the Siwalik deposits explored by Falconer of a still existing species of crocodile among 'many strange and lost mammals'. The South American vertebrates provided a fine example of non-synchronous evolution in different groups (p. 326):

if the Megatherium, Mylodon, Macrauchenia, and Toxodon had been brought to Europe from La Plata, without any information in regard to their geological position, no one would have suspected that they had coexisted with still living sea-shells.

There was nevertheless some evidence of simultaneous change among marine organisms on a worldwide scale. This, however, could be explained by the theory of evolution: forms which had become dominant in the struggle for existence would tend to spread rapidly, until they were replaced by even better-adapted forms.

The main question raised by palaeontology, however, was that of the affinities between extinct and living species. On this point, Darwin could use the results of palaeontologists such as Owen, who had shown 'how extinct animals fall in between existing groups'. Even apparently widely separated groups, such as the ruminants and the pachyderms, which Cuvier had considered the two most distinct orders of mammals, had proved to be connected by fossil forms, which made a revision of classification unavoidable. It was often observed, moreover, that the more ancient forms tended to connect more widely separated groups. There were exceptions in the form of 'living fossils' such as the extant lungfish, but there nevertheless was some truth in the remark. This fact could be explained by the 'theory of descent with modification' (as Darwin called his conception of evolution): divergent evolution from a common original stock could produce the patterns revealed by the fossil record.

A consequence of the theory was that the fauna of any epoch in earth history should be intermediate between that of the preceding epoch and that of the following one. This indeed was the general rule indicated by palaeontology, although there were exceptions, such as those recorded by Falconer among the mastodons and elephants from the Siwaliks. Such exceptions found an easy explanation, however, when it was realised that species had no fixed duration: primitive forms could endure for long periods of time, side by side with more advanced ones.

Darwin did not wish to enter the progressionist controversy, for the simple reason that there was no agreement as to 'what was meant by high and low forms'. The only sense in which more recent forms could be 'higher' than older ones, was in having been more successful in the struggle for life. This superiority, however, could be tested only when two species, or groups of species, were in real competition with each other, and this of course could not be done with extinct faunas or floras.

More important was the interesting fact of the 'succession of the same types within the same area'. This had already impressed Darwin when he excavated fossil mammals in Argentina, and other discoveries had confirmed the generality of the observation: fossil mammals from Australian caves were closely allied to the living marsupials, just as those from the Brazilian caves explored by Lund and Clausen were related to living South American forms. This 'law of succession of type' was not absolute, however, and the fossil record did show that important changes had taken place in the past: marsupials had once been abundant in Europe, faunal resemblances between North and South America had been greater in the past than today, and so on. This fairly complicated pattern of faunal succession again was understandable in terms of 'descent with modification': the fauna of a given period being descended from that of the preceding one, there had to be regional resemblances between the succeeding assemblages in geographically isolated regions. As the earth's surface was subjected to gradual but far-reaching changes, however, intermigration often became possible, and altered the regional character of some faunas. All this did not mean that, for instance, the present edentates of South America were descended from such giants as the ground sloths or the glyptodons; their common ancestors had to be sought among the older and smaller species, and extinction had played a great part in shaping the general aspect of the living fauna.

This question of extinction was also much illuminated by the theory of descent. Catastrophism did not adequately explain the patterns of extinction discovered by palaeontologists, which indicated gradual and non-synchronous disappearances. Groups of organisms obviously had widely different durations, and extinction, on the whole, seemed to be a gradual and lengthy process (although there were exceptions, such as the 'wonderfully sudden' extinction of the ammonites). Again, Darwin went back to his South American observations: the horse tooth he had found there together with large extinct mammals had filled him with astonishment; the recent extinction of the horse in the apparently so favourable South American environment was a mystery. The mystery was more apparent than real, however, when one realised that many species were rare because, for reasons difficult to determine, their conditions of existence were unfavourable to them. As rarity preceded extinction, 'unperceived agencies' could easily render these conditions so unfavourable that a species could become wholly extinct. Although Darwin seemed to be content with it, this was hardly a satisfactory explanation, and the problem of identifying the 'unperceived agencies' is still faced by all palaeontologists trying to explain the extinction of any group.

Darwin's theory did provide a powerful extinction factor, however: competition in the struggle for existence almost inevitably led to the extinction of the 'less-favoured forms'. The appearance of new forms and the disappearance of old ones were thus bound together, and extinction became an integral part of the process of evolution. What would now be called 'mass extinctions', such as that of the ammonites, remained a problem. Darwin's answer involved a *deus ex machina* which he would often use when confronted with embarrassing palaeontological problems: the imperfection of the geological record. If, as he thought, long periods of time had left virtually no geological record, slow extermination might well have taken place during these unrecorded periods. However, he also envisaged that sudden immigration or unusually rapid development of a new group could lead to really rapid extinction of older 'inferior' groups.

Thus, the theory of natural selection could explain all the features of the past history of life which palaeontology had reconstructed, at least in principle. Practical difficulties nevertheless remained, and a major one was that the theory predicted that 'infinitely numerous transitional links' had existed between the many existing and extinct species. Few of these connecting links had actually been found, however, and in many cases the fossil record suggested a sudden appearance of whole groups of species. To counter this serious objection, Darwin resorted to the imperfection of the record, to which he devoted a whole chapter. An important point was that our knowledge of geological succession on a worldwide scale was far from complete, and that palaeontological collections were still very poor, both because many parts of the world were still unexplored (from a palaeontological point of view) and because many organisms had not been preserved as fossils. Of course, it could still be hoped that further research would fill many gaps in our knowledge, but Darwin felt it necessary to take a rather pessimistic stand in this regard. To him, the very nature of the geological record made it unavoidable that long periods of time had left no record at all. Even within a Lyellian, uniformitarian framework, the destructive role of erosion had been of prime importance. All the richly fossiliferous formations had accumulated during periods of subsidence, but the slow 'oscillations of level' which had affected all parts of the world had brought about repeated periods of widespread 'degradation', during which many older deposits had been completely destroyed. Because of this, the fossil record was bound to be 'intermittent', interrupted by many gaps, which prevented any reconstruction of a complete and continuous succession. No wonder then that very few 'transitional links' had been preserved, and the absence of the corresponding fossils made it extremely difficult to recognise evolutionary lineages. Although palaeontology had revealed some forms which 'made the intervals between some few groups less wide than they otherwise would have been', the finer gradations between species remained virtually unknown, and there was relatively little hope of ever being able to bridge such gaps, because of the imperfection of the record.

Although partly justified, Darwin's pessimism did not prevent palaeontologists who had rallied to his views searching for those elusive 'missing links'. They were encouraged in this search by some spectacular discoveries, especially in the field of vertebrate palaeontology which, shortly after the publication of the *Origin of Species*, clearly demonstrated that intermediate forms could indeed be found.

One of the most striking of these discoveries was undoubtedly that of the earliest known bird, *Archaeopteryx* (Wellnhofer, 1983). As early as 1855, a poorly preserved bird skeleton had been found in the lithographic limestones of the Altmühl valley in Bavaria; it had been described by Hermann von Meyer, who had failed to recognise its true nature and had thought it a pterosaur (it was not until 1970 that the specimen was finally correctly identified by the American palaeontologist John Ostrom). In 1860 quarrymen found an isolated feather in a limestone quarry near Solnhofen, in the same region of Germany. Hermann von Meyer understood the importance of this find, which clearly demonstrated that birds had been in existence in the Jurassic, much earlier than usually admitted; he coined the name *Archaeopteryx lithographica* for the specimen. In 1861 an even more exciting discovery was made in another quarry in the vicinity of Solnhofen: this was the skeleton of a small vertebrate which in many respects was reminiscent of a reptile, but was surrounded by distinct feather impressions. The rather complicated story of this famous specimen shows that recognition of the scientific importance of such fossils resulted in a considerable rise of their commercial value. Carl Häberlein, a physician in the small town of Pappenheim, a short distance from Solnhofen, was able to obtain the specimen soon after it was discovered. Häberlein was the owner of a large collection of fossils from the lithographic limestones, many of them provided by poor quarrymen who paid for his services with fossils (Viohl, 1985). Häberlein had a large family to care for, and he soon realised that the feathered fossil was an extremely valuable specimen which could be sold at a high price. A fossil collector from Hanover, O.J. Witte, saw the fossil and tried to have it purchased by the State Palaeontological Collection in Munich. The curator of the collection was Andreas Wagner, who became interested in the fossil and sent his assistant Albert Oppel to examine it. Häberlein, however, would not allow visitors to make drawings or write descriptions of the precious specimen, for fear that it would lose some of its value if a scientific description was published. Oppel is said to have spent several hours looking at the fossil, after which he returned to Munich and there made a remarkably accurate drawing of it from memory. However, the drawing, which is still preserved in Munich, is so accurate that this version is hardly credible, and it seems more likely that he managed to make a sketch of the skeleton during his stay in Pappenheim (Wellnhofer, 1985). In any case, Oppel's report was used by Wagner as the basis for a description of the mysterious animal. Wagner was a rather old-fashioned 'biblical' geologist, who even published computations about the size of Noah's Ark (Wagner, 1858), and he was not prepared to admit the intermediate nature of the fossil. He therefore described it as an unusual reptile, which he called *Griphosaurus* (Wagner, 1861), and strongly opposed any Darwinian interpretation of it (Ostrom, 1985). Meanwhile, Carl Häberlein had offered the specimen for sale, together with the rest of his collection, to the British Museum. Richard Owen was then superintendent of the Natural History Departments, with Robert George Waterhouse as keeper of mineralogy and geology. Both men were anxious to acquire Häberlein's collection, and in 1862 Waterhouse travelled to Pappenheim to examine it. The collection was finally purchased by the British Museum for £700 (which sum Häberlein used, it is said, as a dowry for his daughter). At the end of 1862 the Häberlein collection was sent to London, and in 1863 Owen published a description of the 'London' *Archaeopteryx*, which he interpreted as a primitive bird. Although in many respects *Archaeopteryx* was indeed a bird, several peculiar features (such as the toothed jaws, the long tail and the unfused metacarpals) indicated remarkable resemblances with reptiles, and the 'intermediate' nature of this earliest bird was soon accepted by most palaeontologists. Thomas Huxley, for instance, an early convert to Darwinism, thought that *Archaeopteryx* exhibited many similarities with the very small dinosaur *Compsognathus*, then known from a single skeleton also described by Wagner from the lithographic limestones of Bavaria. Although there was no general agreement as to which group of reptiles was most closely related to birds, *Archaeopteryx* quickly became the archetypal 'missing (or rather, no longer missing) link' between two major groups of vertebrates, and its fame soon extended beyond scientific circles (as evidenced, for instance, by the use of *Archaeopteryx* as a character in a play by the French author Alfred Jarry in 1897–8; see Buffetaut, 1985).

When a second and remarkably well-preserved *Archaeopteryx* skeleton was found near Eichstätt, in the Altmühl valley, in 1877, it was acquired by Carl Häberlein's son Ernst, who, following his father's example, offered it for sale together with his fossil collection. Interest in *Archaeopteryx* had, if anything, increased since the discovery of the 'London' specimen; this had resulted in inflation, and Ernst Häberlein asked a very high price (36,000 Marks) for the new find. The purchase of the new *Archaeopteryx* became a question of national pride in Germany, and after painstaking negotiations (which resulted in the price being lowered to 20,000 Marks), the Prussian State finally acquired the specimen for the Humboldt University in Berlin with money which had been loaned by the famous industrialist Werner von Siemens (Wellnhofer, 1983).

Not all 'intermediate' fossil vertebrates were as spectacular and rare as *Archaeopteryx*; it was soon realised that more 'ordinary' fossils could also provide valuable evidence in favour of evolution. In the 1870s, for instance, the Mesozoic crocodilians which had led Geoffroy Saint-Hilaire to early evolutionary speculations in the 1820s received renewed attention from Thomas Huxley, Darwin's 'bulldog' and one of the first to interpret (or reinterpret) fossil vertebrates in the light of Darwin's theory. Fossil crocodilians had been found in fairly large numbers in the Jurassic rocks of France, England and Germany, and detailed anatomical studies had been conducted on some exceptionally well-preserved specimens by J.A. Eudes-Deslongchamps and his son Eugène in Caen. Moreover, even older crocodile-like creatures had been described by German palaeontologists (among them Hermann von Meyer) from the Upper Triassic of Germany. It seemed that the transformations of such important anatomical features as the secondary palate permitted a reconstruction of a morphological series leading from primitive Triassic forms to the modern crocodiles, via the intermediate Jurassic forms. This view was presented by Huxley in a paper published in 1875, in which he proposed a threefold division of the order Crocodilia into the suborders Parasuchia, Mesosuchia and Eusuchia (a classification which is still partly used today, although it is now recognised that the Triassic Parasuchia, or phytosaurs, are not ancestral to crocodiles, and that the resemblances are due to convergent evolution in similarly adapted groups). Huxley's morphological series could be arranged chronologically according to the geological age of the three suborders; this showed that the Parasuchia were older than the Mesosuchia, and that the Eusuchia were the last to appear, as would be expected if the morphological series reflected the evolution of the crocodilian lineage. Huxley's conclusion was unequivocal (Huxley, 1875, p. 431):

Thus the facts relating to the modifications which the Crocodilian type has undergone since its earliest known appearance, are exactly accordant with what is required by the theory of evolution; and the case of the Crocodiles is as cogent evidence of the actual occurrence of evolution as that of the Horses.

Further confirmation of Huxley's interpretation was provided by the preparation of a fragmentary crocodilian skull from the Wealden of the Isle of Wight, which revealed internal nares in a position intermediate between that of the Jurassic mesosuchians and that of the Tertiary eusuchians. An intermediate morphology thus again corresponded to an intermediate geological age.

Huxley's interpretation of the crocodilian fossil record was hailed as a brilliant exercise in evolutionary palaeontology, and the transformations of crocodilian morphology were often used as an especially clear and convincing palaeontological demonstration of the reality of evolution. Interestingly, in 1878 Richard Owen himself delivered a paper at the Geological Society of London 'On the influence of the advent of a higher form of life in modifying the structure of an older and lower form', in which he interpreted changes in crocodilian morphology during the Mesozoic as a consequence of the appearance and expansion of mammals. According to his interpretation, changes in the vertebrae, scutes and skull of crocodilians were linked to the decline of large reptilian predators (which allowed the dermal armour to regress) and to the appearance of terrestrial mammalian preys, which largely replaced fish in the crocodilian diet (this supposedly made a different mode of locomotion and predation necessary, and resulted in changes in the jaws, palate, limbs and vertebrae). Owen's conception of changes in the faunal environment leading to morphological transformation did sound like an evolutionary interpretation, and in the discussion which followed the presentation of his paper, several participants remarked upon this. As a Dr Meryon felt 'sorry that the author had not gone to the full length in regard to evolution', Owen answered that 'his paper was teleological rather than evolutionist': his aim had been to examine how changing structures could provide an advantage in the capture of terrestrial prey. Nevertheless, he did suggest a possible Lamarckian process through which these changes could be effected, and he concluded that 'it may reasonably be expected that many more of these "missing links" will be found'. It seems that, despite his continuing opposition to natural selection, in the 1870s Owen's thinking had come very close to an acceptance of some kind of evolution.

The acceptance of evolution, not necessarily accompanied by an acceptance of natural selection, by British palaeontologists was emulated on the Continent. There, as in Britain, strong opposition had sometimes to be vanquished. In France, for instance, Albert Gaudry (1827–1908) at first found it hard to propagate his evolutionary interpretations of fossil mammals. In the 1860s he had been asked to teach palaeontology at the Sorbonne, but his unorthodox views soon led the dean of the faculty, without even informing Gaudry, to delete his courses from the official programme (Thevenin, 1910). Gaudry had started his career as a geologist at the Natural History Museum in Paris. In 1853 he went to Cyprus on a geological and agronomical mission, and on his way back to France he stopped in Athens, where he heard about the discovery of Tertiary mammals at Pikermi, in Attica. Fossil bones had first been found there by the English archaeologist George Finlay, who started excavations on a small scale with the ornithologist Lindermayer (Abel, 1927). In 1838, after the Bavarian prince Otto von Wittelsbach had become King of Greece, a Bavarian soldier found more bones at Pikermi. He thought that the calcite crystals in the cavities of some of them were diamonds, and took his finds back to Munich. There, they came into the hands of Andreas Wagner, who described them in 1839. Among the most interesting specimens was the skull of a monkey (in 1839 Tertiary monkeys were still a novelty). After that, excavations were started again first in 1848 by Lindermayer, and then on a larger scale by Roth in 1852 and 1853; the remains thus discovered were described by Roth and Wagner in Munich. In 1853 the Greek government became interested in the Pikermi fossils and specimens were collected for the University of Athens by Mitzopoulos. Gaudry therefore was not the first to collect fossils in the Miocene deposits of Pikermi (and not the last either, since German, French, Austrian and Greek parties excavated there after him), but he interpreted the results of his excavations in a new way. In 1855–6, having obtained funds from the French Academy of Sciences, Gaudry spent a first field season at Pikermi (under the protection of soldiers because bandits roamed the countryside and kidnapped travellers). The excavations were continued in 1860, when he spent seven months there with his young wife. Despite bandits and fevers, Gaudry immensely enjoyed his field work in Greece, and he has left enthusiastic descriptions of his stays at Pikermi and of the bucolic feasts which were given there when an especially interesting specimen was discovered (Gaudry, 1888). Thousands of bones were collected and shipped to Paris. After several preliminary notes, Gaudry published his first great palaeontological monograph, *Animaux fossiles et géologie de l'Attique*, between 1862 and 1867. His excavations and those of his predecessors had revealed a rich and varied late Miocene fauna, comprising rhinoceroses, monkeys, sabre-toothed cats, hyaenas, abundant antelopes, the three-toed equid *Hipparion*, primitive giraffes and proboscideans. Gaudry grew lyrical when he described the Miocene fauna of Greece (1888, p. 89): 'the most majestic of all animals was the *Dinotherium*. How beautiful it must have been, when it came forth, escorted by the mastodon with breast-like teeth and the mastodon with tapiroid teeth.'

The peculiar proboscidean *Deinotherium* had been the subject of much controversy. Cuvier had misidentified isolated teeth of it as those of a giant tapir. In the 1830s good cranial material had been found at Eppelsheim in Germany, and described by Kaup and Klipstein. However, the postcranial skeleton remained unknown, and the bizarre downturned tusks of the lower jaw were given strange interpretations. Buckland had reconstructed *Deinotherium* as a somewhat walrus-like aquatic animal, and many early reconstructions depicted it in a lying position, to avoid showing the unknown limbs (Abel, 1925). Gaudry's finds provided much better information and demonstrated that it was indeed a proboscidean. He seems to have been strongly impressed by the imposing stature of this animal (Gaudry, 1888, p. 103): 'This giant of the ancient world, both powerful and peaceful, which none had to fear, which was respected by all, was truly the personification of the calm and majestic nature of geological times.'

The fossils from Pikermi were not simply the basis for picturesque reconstructions of the life of Miocene savannas; they also led Gaudry to evolutionary interpretations which went far beyond what had so far been done in France in this field. The Pikermi fauna revealed many of these intermediate forms, the lack of which had seemed to be an obstacle to the theory of evolution. In Gaudry's words (1888, p. 144):

> Thanks to the palaeontological researches which are everywhere conducted, beings of which we did not understand the place in the economy of the organic world are revealed to us as links in chains which themselves are connected; one finds transitions from order to order, from family to family, from genus to genus, from species to species.

The monkey from Pikermi, *Mesopithecus pentelici*, for instance, was intermediate between *Macaca* and *Semnopithecus*; the Pikermi rhinoceros had a strengthened nasal region which placed it between the hornless *Aceratherium*, of the Miocene, and the later forms; *Hipparion* was intermediate in the structure of its three-toed feet between *Anchitherium* and the modern horse, etc. 'Intermediate types' could thus be found in almost all groups of mammals discovered at Pikermi. This, however, was not a particularity of the Pikermi fauna: fossils with intermediate characteristics could be found in all localities. This important fact led Gaudry to more 'philosophical' considerations concerning the transformation of species. Faunal renewal now turned out to be a continuous phenomenon: d'Orbigny's twenty-seven phases of faunal change could no longer be accepted. One had to choose between continuous creation and transformation. Transformation was obviously simpler to admit, but this in itself was not sufficient, as 'everything is equally easy to the Creator'. However, the many intermediate forms found in the fossil record provided strong evidence in favour of transformation, despite the fact that many gaps remained, and that some fossils were intermediate in some respects only, which made it impossible to include them in a true genealogy. Many uncertainties remained, but Gaudry had no doubt that the progress of palaeontological research would provide more and more data in support of evolution. When it came to the mechanism of evolution, Gaudry became much more diffident. He insisted on the need to separate the question of whether transformations had taken place (which could be solved by a careful study of fossils) from that of the means by which they had been effected. According to Gaudry, Darwin's merit had been to show that transformation occurred, but serious objections could be made to his hypotheses on the mechanism of evolution. Gaudry had read Darwin's *Origin of*

Species 'with passionate admiration', he had 'savoured it slowly, as one drinks a delicious liquor' (Gaudry, 1888, p. 32) — and he had received encouragement from Darwin himself when he started to publish on evolutionary palaeontology. Nevertheless, he admitted that he had 'always been very far from Darwin's philosophical ideas in some respects' (1888, p. 32). The reason for this was clear: Gaudry's strong religious feelings made it difficult for him to accept Darwin's mechanistic vision of an evolution based on chance and natural selection. Evolution was acceptable to him because it revealed a unity in the organic world which was the mark of divine works. He did not draw a sharp boundary between creation and evolution. The question was not whether or not God had created, it was whether creation had been through formation or through transformation. Gaudry's rather idyllic vision of the Pikermi fauna was in complete agreement with his conception of a harmonious divine creation through evolution (Gaudry, 1888, p. 168):

> Whatever the way in which animals have been renewed, what is certain is that no modification has been the result of chance. My researches have shown that, during geological times, Greece was not the theatre of struggle and disorder; everything there was harmoniously set. If we recognise that organised beings have little by little been transformed, we shall regard them as plastic substances which an artist has been pleased to knead during the immense course of ages, lengthening here, broadening or diminishing there, as the sculptor, with a piece of clay, produces a thousand forms, following the impulse of his genius. But we shall not doubt that the artist was the Creator himself, for each transformation has borne a reflection of his infinite beauty.

Gaudry was fortified in his evolutionist convictions by his further researches on fossil vertebrates. He felt the need to demonstrate that fossil species had been variable enough for transspecific evolution to be possible, and he thought that the study of a fauna contemporaneous with that of Pikermi could provide important data in this respect by showing whether variations were observable. He chose a Miocene locality in the Léberon (or Lubéron) mountains of south-eastern France, where he collected abundant bones of mammals very similar to those from Pikermi. He interpreted many of the small morphological differences he noticed as intraspecific variations, which were not sufficient to warrant the creation of new species. This widespread kind of variation was the first step of evolutionary divergence. Another result of the comparative study of the Léberon and Pikermi faunas was the recognition of the importance of migration: the faunal differences between geological stages or substages often could be explained by the immigration of organisms from another region.

Although he always remained interested in fossil mammals (his last publications were on South American Tertiary mammals), Gaudry broadened the scope of his investigations when he studied the Permian amphibians from the Autun region of central France. These imperfectly ossified quadrupeds were 'young' in an evolutionary sense — and the juvenile features of living vertebrates resembled the early stages of vertebrate evolution. This confirmed some of the old observations of Agassiz, as well as Haeckel's much more progressive ideas about the parallelism between ontogeny and phylogeny. In other respects, however, the characteristics of Permian tetrapods were not in agreement with some theoretical conceptions: they could not be reconciled with Owen's archetypal theory, for instance.

Gaudry's evolutionary interpretation of the fossil record was brought together in his three-volume work *Les enchaînements du monde animal dans les temps géologiques*, published between 1878 and 1890. This was an ambitious depiction of the course of evolution, from Palaeozoic invertebrates to Cenozoic mammals, and these well-written and profusely illustrated volumes were widely read at the time. They were followed in 1896 by an *Essai de paléontologie philosophique*, in which Gaudry summarised his conclusions on the evolution of life as revealed by the fossil record. The most obvious feature of this evolution was that it had been progressive: as living beings became more diversified, they also became more active, more sensitive and more intelligent. The fact that evolution was not a random process had practical applications: once the general direction of evolutionary change within a group had been determined, it became possible to use the 'evolutionary stage' of a fossil as a clue to its geological age. Despite some irregularities in the course of evolution, this, according to Gaudry, made biostratigraphy a more efficient tool for geology than it had been when mere associations of fossils had been used to correlate strata.

In his conclusion Gaudry examined more purely philosophical

questions. He had no doubt that the history of life had been directed by a divine plan, and that palaeontology was nothing but the study of this plan. Man was the ultimate goal of this plan, toward which progressive evolution was aimed. The reality of extinction could not be denied, but this was an unpleasant phenomenon on which Gaudry did not wish to dwell. There was no place for competition and selection in Gaudry's divinely ordained completely harmonious world in which the purpose of carnivores was to relieve herbivores of their sufferings when they were threatened with starvation because of overpopulation . . . His idyllic view of evolution was the very opposite of Tennyson's 'nature red in tooth and claw' and of Darwin's survival of the fittest (Gaudry, 1896, p. 30):

It is said that the living beings of the various geological ages have fought each other and that the strong have defeated the weak, that the battlefield was left to the fittest; thus progress would be the result of the fights and sufferings of the past. This idea is not borne out by palaeontology. The history of the living world shows us an evolution in which everything is combined as in the successive transformation of a seed·which becomes a magnificent tree covered with flowers and fruit, or an egg which changes into a complicated and charming creature.

His opinion had not changed since the 1860s, when he wrote about Pikermi (Gaudry, 1867, p. 19): 'Thus, there was no competition for life, everything was harmonious, and He who today regulates the distribution of living beings, regulated it in the same way in past ages.'

A very different approach was used by the Russian Vladimir Kovalevskii (1842–83), who was also one of the first in Europe, with Gaudry and the Swiss Rütimeyer, to interpret fossil vertebrates in an evolutionary way (Todes, 1978). In contrast to Gaudry, Kovalevskii had become an evolutionist (and a Darwinist) before he became interested in palaeontology. In the 1860s he had been a political activist and had led an adventurous life. After visiting Germany, France and England (where he met Darwin at Down in 1866), he returned to Russia and translated into Russian some of Darwin's and Huxley's works. In 1868 he married a gifted mathematician, Sofia Korvin-Krukovskaia, and they left for western Europe to study. Kovalevskii developed an interest in vertebrate palaeontology after reading Cuvier's works. Despite the Prussian blockade and the Commune, he managed to visit Paris several times in 1870 and 1871, and met Gervais and Gaudry there. He worked on fossil mammals in the National Museum and at the Ecole des Mines. Lartet had just died and Kovalevskii obtained permission from Gervais and Gaudry to study the *Anchitherium* remains which were kept in the Parisian collections. He interpreted this form as an intermediate in a lineage which led from *Palaeotherium* to *Equus* through *Anchitherium, Merychippus* (a then newly discovered North American form) and *Hipparion*. This phylogenetic reconstruction was published in 1873 in a monograph which was the first in a series of four, which make up the whole of Kovalevskii's palaeontological work. His aim was to provide an account of the evolutionary development of all the ungulates, and his point of view was unambiguously Darwinian, as testified by his 1873 monograph on *Anthracotherium*, which was dedicated to Charles Darwin. Contrary to Gaudry, Kovalevskii thought that palaeontology could provide valuable evidence on the *process* of evolution. In this way, he explained toe-reduction in ungulate evolution as a result of advantageous changes brought about by adaptation and competition. Environmental change was supposed to have played an important part in this process: the development of open plains and steppes had been favourable to the expansion of ruminants. Kovalevskii's work attracted much attention among palaeontologists both in England and on the Continent (Huxley and Gaudry were among those who expressed appreciation of it). In Russia, however, scientists were mostly uninterested. Kovalevskii had gained a doctorate in Jena, with Ernst Haeckel among the examiners, but when he returned to Russia in 1873 and tried to obtain a degree in Odessa, he failed because of the hostility of the examiners. Eventually he managed to obtain a master's degree in St Petersburg in 1875, but he could not find a position as a palaeontologist. He later decided to stop doing scientific work for a few years and engaged in several unsuccessful business ventures. He worked for some time for an oil company before finally obtaining a position at Moscow University in 1881. He was not a very successful teacher, and his financial position deteriorated because of the difficulties of the oil company. In 1882 he had the opportunity to meet Cope and

Marsh during a business trip to the United States. Kovalevskii finally became severely depressed because of financial and marital problems, and committed suicide in 1883. Despite the brevity of his scientific career, his work exerted a strong methodological influence on other palaeontologists (among them the Belgian Louis Dollo, the Austrian Othenio Abel and the American Henry Fairfield Osborn), and he was hailed as one of the most influential of the late nineteenth-century evolutionary palaeontologists.

It was thus in Europe that vertebrate palaeontology was first envisaged within an evolutionary framework. However, sensational discoveries of fossil vertebrates in North America were soon to bring American scientists to the forefront of research in this field.

7

Nineteenth-century Vertebrate Palaeontology in the United States

As mentioned in Chapter 3, the discovery of large fossil bones in North America during the eighteenth century played an important part in the early development of vertebrate palaeontology. However, most of the descriptive and speculative work on these fossils was done by European scientists, and it took a long time before North American palaeontologists could exert an influence comparable to that of their British, French or German colleagues. Nevertheless, by the end of the eighteenth century the study of fossils had become a fashionable pursuit among the American *intelligentsia*, as shown by the case of Thomas Jefferson, second president of the United States (Simpson, 1942). Jefferson, as president of the American Philosophical Society from 1797 to 1814, encouraged the study of vertebrate palaeontology, and although he had no time for field work, did his best to obtain as many specimens as possible. He had bones of mastodon, mammoth, bison, musk-ox and deer collected at Big Bone Lick by Captain Clark in 1807, and for a time a room at the White House was set aside to accommodate the collection. Jefferson's main written contribution to vertebrate palaeontology was a publication in 1799 'on the discovery of certain bones of a quadruped of the clawed kind in the western parts of Virginia'. The bones had been found by workmen excavating the floor of a cave for saltpetre. Some of them had been dispersed as curiosities, but Jefferson managed to obtain several specimens. They were limb bones of an unknown mammal, and the most striking of them were large claws. Jefferson thought that they had belonged to a big cat, three times the size of a lion, for which he used the name 'megalonyx' (meaning 'large claw'). After Jefferson had presented the bones to the American Philosophical Society, they were studied in a much more competent way by the anatomist Caspar Wistar, who brilliantly reconstructed the functional anatomy of the foot of *Megalonyx*, and recognised that it resembled the sloth but was different from the newly discovered South American *Megatherium*. As to Jefferson, he apparently never totally abandoned the idea that the Virginia animal may have been some sort of giant cat. Moreover, he never accepted that species could become extinct. In accordance with the old views on the perfection of creation, he believed that 'if one link in nature's chain might be lost, another and another might be lost, till the whole system of things should evanish by piecemeal' (Jefferson, 1799, pp. 255–6). As late as 1825, one year before his death and long after Cuvier had proved the extinction of many animal species, he wrote that 'such is the economy of nature, that no instance can be produced of her having permitted any one race of her animals to become extinct, of her having formed any link in her great work so weak as to be broken' (Simpson, 1984). Jefferson expected the mastodon and the megalonyx to be still living in the wildernesses of western North America, and it is said that when Lewis and Clark set out on their famous expedition to the Far West, they had instructions to search for living representatives of these fossil beasts. They did not find living mastodons, but they did discover fossil bones, including a fish jaw which was later described by Harlan as *Saurocephalus lanciformis*, and came from a cave close to the Missouri River. Their other finds, which were not collected, may have been bones of Mesozoic reptiles (Simpson, 1942).

However, and despite several isolated finds (listed by Simpson, 1942), the palaeontological exploration of the western United States did not really start until several decades later. During the first half of the nineteenth century most of the major discoveries were made in the eastern part of the country, and many of the fossils found there belonged to the large Pleistocene fauna. The American mastodon was still prominent among them. Thus, in 1799 bones were discovered in a peat-bog on the farm of John Masten in Orange County, New York State. With a party of more than a hundred neighbours, Masten excavated the site 'with more energy than care', to use Simpson's words, some of the men indulging in spirits to the point of becoming 'impatient and unruly', with the result that many bones were damaged (Simpson, 1942). Two years later, Charles Willson Peale (1741–

1827) heard about the find. Peale, an artist, had gathered a miscellaneous collection of paintings and natural history objects at his Philadelphia house, and had converted it into a museum which was eventually transferred to the hall of the American Philosophical Society. Peale was anxious to augment the number of vertebrate fossils in the museum, and John Masten's find was promising. Digging rights were obtained for $100, and a systematic excavation was started with financial support from the American Philosophical Society. As the pit was flooded, complex machinery had to be installed to drain it, and a ship's pump had to be brought from New York. The result was an elaborate contraption, depicted by Rembrandt Peale, Charles's son, in a famous picture (Howard, 1975). The excavation attracted much attention and large crowds gathered around the pit, in which most of the skeleton of a mastodon was eventually recovered. Such essential parts as the lower jaw were still missing, however, and excavations were started at other sites in the vicinity to try to find them. Finally, a second skeleton was found on the farm of a Peter Millspaw, where a few bones had been discovered in 1795. The skeleton was less complete than the first one, but it included the lower jaw. Rembrandt Peale (1803) has left an enthusiastic account of its discovery:

> The unconscious woods echoed with repeated huzzas, which could not have been more animated if every tree had partici-pated in the joy. 'Gracious God, what a jaw! how many animals have been crushed beneath it!' was the exclamation of all: a fresh supply of grog went round. . .

Obviously, the old conception of the carnivorous mastodon was not yet dead. The bones were wrapped in canvas and taken to Philadelphia, where a painstaking process of reconstruction took place before the best-preserved skeleton could be mounted and displayed in the Philosophical Society's hall. This was apparently the second fossil vertebrate skeleton ever to be mounted (the first one being the Luján *Megatherium* mounted in Madrid by Brú).

Older fossil vertebrates were also found in the eastern United States in the early years of the nineteenth century. Among them were remains of fishes and reptiles from the late Cretaceous New Jersey Greensand, which were described by local physician-naturalists such as Richard Harlan in Philadelphia and James

Ellsworth De Kay in New York. Harlan, who had realised the importance of reconstructing the chronological succession of fossils through geological time, has already been mentioned in connection with the discovery of the early whale *Basilosaurus* and the use made of it in France and England in the controversy about the Stonesfield mammals. *Basilosaurus* vertebrae had originally been reported from Eocene marls in Louisiana in 1832 (Kellogg, 1936), and described by Harlan as belonging to a reptile. Additional bones and teeth were found in 1834 and 1835 on a plantation in Alabama, and this material enabled Owen to demonstrate conclusively that *Basilosaurus* was in fact a mammal. More remains were discovered in the 1840s in Alabama, and they attracted the attention of Albert Koch, a German collector, who had little scientific ability but knew how to collect fossils and how to derive financial benefits from public interest in them. In 1832 he had interpreted some poorly preserved mastodon bones as those of a new and gigantic creature which he called the *Missourium*. He eventually mounted a huge and fantastic skeleton with the remains of several mastodons, and exhibited it as that of the *Missourium*, which, according to him, was none other than the Biblical Leviathan (Simpson, 1942). The bones were purchased by the British Museum in 1844, and some of them were re-mounted as a more normal mastodon skeleton (Stearn, 1981). Koch's treatment of the archaeocete bones he collected in Alabama in 1845 was hardly better: with bones from at least five individuals, he reconstructed a 114-foot (35m)-long animal, which he dubbed *Hydrargos sillimannii* and exhibited as a sea-serpent in New York City. The composite skeleton was later shipped to Germany and exhibited in several European cities. German scientists then had an opportunity to study it. In 1847 the King of Prussia purchased Koch's *Hydrargos* for the Berlin Anatomical Museum, and in the same year Johannes Müller was finally able to demonstrate that the skeleton was a composite one and gave an adequate description of the archaeocete remains on which it was based (Kellogg, 1936).

Although they attracted much attention at the time, Koch's fantastic reconstructions and interpretations were not really representative of the work done on North American fossil vertebrates. Most of it was much more serious, even though some fossils did prove difficult to interpret. An example of this is provided by Edward Hitchcock's work on the fossil footprints of the Connecticut valley. As early as 1800, a three-toed fossil footprint had been found near South Hadley, Massachusetts, by Pliny Moody, a student at Williams College (Colbert, 1970). The footprint was bird-like, and Moody supposed that it had been left by Noah's raven. In 1835 more footprints were found by a Mr Draper, who thought they were turkey tracks. The village physician heard about them and notified Edward Hitchcock, who was then president of Amherst College as well as State Geologist of Massachusetts. Hitchcock became very interested in these mysterious footprints and devoted much of his career to their study (Colbert, 1968). He eventually assembled a large collection of Triassic footprints at Amherst College, which was described in 1858 in his *Ichnology of New England*. Most of the footprints were those of dinosaurs, but Hitchcock thought that they had been made by large, sometimes gigantic birds, although he suggested that some had perhaps been left by amphibians, reptiles and even marsupials. He thus pictured to himself 'an apterous bird, some twelve or fifteen feet [3–4m] high', living in flocks, a bipedal creature with a foot and heel nearly two feet (0.6m) long, and a host of other peculiar birds or bird-like animals. The occurrence of birds in the Triassic period, much earlier than was indicated by bony remains, was of course used by anti-progressionists in the controversy against the supporters of a successive appearance of increasingly developed vertebrate types. Hitchcock's *Ornithichnites* were thought to demonstrate an early origin of this class of 'higher' vertebrates, just as *Chirotherium* was supposed to prove the existence of mammals at the dawn of the Mesozoic era. The discovery of *Archaeopteryx* did not alter Hitchcock's opinion; it even fortified him in his belief that the Connecticut valley footprints had been made by early birds (Colbert, 1968). It was only after his death that the dinosaurian origin of most of them became generally accepted, when the bird-like structure of some dinosaurs began to be recognised by such palaeontologists as Thomas Huxley.

Not only dinosaur footprints, but dinosaur bones too could be found in the Mesozoic strata of the eastern United States, and the identification of the first North American dinosaur skeleton by Joseph Leidy was a landmark in the study of these reptiles. Leidy (1823–91) had studied medicine at the University of Pennsylvania, but failed when he tried to become a practitioner. He turned to an academic career, and finally became professor at the

University of Pennsylvania. In 1858 he heard of an exciting discovery which had been made at Haddonfield in New Jersey, not far from Philadelphia. A Mr Foulke had gone to spend the summer there and had heard that twenty years before, his neighbour Mr Hopkins had discovered large vertebrae in Cretaceous marls on his farm. The fossils had been taken away as souvenirs by various visitors. Foulke obtained permission to have the site excavated by workmen, who came across a number of large bones. Leidy visited the locality, and later described the fossils under the name of *Hadrosaurus foulkii*. The material consisted of 9 teeth, a fragmentary lower jaw, 28 vertebrae, bones of the front and hind limbs, and pelvic bones. This was enough to get an impression of the general appearance of the skeleton, and Leidy was led to the conclusion that *Hadrosaurus*, with its relatively short front limbs, may have been rather kangaroo-like in posture, although it also could walk on all fours. This was a radical departure from Owen's conception of dinosaurs as large heavy-limbed quadrupeds, and it marked the beginning of a reappraisal of dinosaurian morphology and behaviour. Some ten years after the discovery of *Hadrosaurus*, the new interpretation materialised in a tangible fashion when Waterhouse Hawkins, the sculptor who had produced the Crystal Palace reconstructions under Owen's supervision, started to work on a vast project of a 'Palaeozoic Museum', to be erected in Central Park, in which life-size reconstructions of extinct animals were to be exhibited (Colbert, 1968). To judge from drawings made at the time, the restored *Hadrosaurus* was to be a bipedal creature rather similar to a modern reconstruction of an ornithopod dinosaur. Unfortunately, the Palaeozoic Museum was never completed, because of political machinations in New York City, and all that remains of Waterhouse Hawkins's American reconstructions are drawings and paintings.

Actually, the Haddonfield *Hadrosaurus* was not the first dinosaur specimen to have been described by Leidy. In 1855 he had received a few fossil teeth collected by Dr Ferdinand Vandiveer Hayden in Cretaceous rocks near the confluence of the Judith River with the Missouri (in what is now Montana). Leidy recognised some of them as belonging to a herbivorous reptile resembling *Iguanodon*, which he called *Trachodon*; others were reminiscent of the European *Megalosaurus*. This was the first record of dinosaurs in North America, as well as an important step in the palaeontological exploration of the

127

'Western Territories'. Although he had few opportunities to go west himself, and usually had to be content with fossils collected by others, Leidy played an important part in the discovery of the palaeontological treasures of the western United States. As mentioned above, Lewis and Clark had already found some vertebrate remains during their famous expedition, and other explorers after them had also made interesting discoveries. Among the most promising regions were the 'Mauvaises Terres' (thus named by French explorers), or Badlands, of Nebraska, 'an area of country extending along the foot of the Black Hills'. This deeply eroded, sparsely vegetated area was a nuisance to travellers, but, as remarked in a report quoted by Leidy (1853, p. 12):

> The drooping spirits of the scorched geologist are not permitted, however, to flag. The fossil treasures of the way well repay its sultriness and fatigue. At every step, objects of the highest interest present themselves. Embedded in the debris, lie strewn, in the greatest profusion, relics of extinct animals.

A Dr Hiram Prout of Saint Louis first brought the Badlands fossils to the notice of the scientific world in 1847, with a description of a maxilla which he attributed to *Palaeotherium*. In the late 1840s and early 1850s, several explorers collected fossils in the 'Mauvaises Terres', and their finds were eventually described by Leidy in his monograph on 'The ancient fauna of Nebraska', published by the Smithsonian Institute in 1853. This ancient fauna, which he thought was of Eocene age (it is now known to be Oligocene), included many forms new to science, such as the large perissodactyl *Titanotherium* or the abundant artiodactyls which Leidy placed in the genus *Oreodon*. A sabre-toothed cat was also present, as well as large land tortoises. Leidy gave a factual and detailed description of these animals, and did not indulge in speculations about their significance, but the importance of the palaeontological discoveries made in the 'Mauvaises Terres' was clear: the American West was a potential source of palaeontological riches. This was confirmed by further discoveries of fossil mammals in the same region, which enabled Leidy to enlarge his 1853 work, and to publish in 1869 a 472-page volume entitled *The Extinct Mammalian Fauna of Dakota and Nebraska*. By 1869 the number of pre-Pleistocene mammal species known from North America amounted to eighty-four, most of them from the Dakota and Nebraska Badlands (Lanham, 1973).

In the 1860s the Civil War brought scientific exploration of the West to a standstill. After peace returned, however, the scientific surveys of the western territories started again, on a larger scale (Faul and Faul, 1983). The various expeditions led by men such as Ferdinand Hayden, John Wesley Powell, Clarence King and George Wheeler in the 1870s confirmed that palaeontological treasures waited to be discovered in the largely unexplored deserts and mountains of western North America. The construction of the transcontinental railway lines also resulted in important discoveries. At first, Joseph Leidy was the main beneficiary of the new discoveries, but very soon the competition of two newcomers on the scientific scene, Edward Drinker Cope and Othniel Charles Marsh, was felt, and Leidy decided to retire from the field of vertebrate palaeontology, and turned instead to the study of protozoans. One of the reasons for this was of a financial nature: Cope and Marsh could afford to pay for fossils, which Leidy could not. Another reason was that the feud which soon developed between Cope and Marsh was not to Leidy's taste. This quarrel, which soon extended well beyond the bounds of scientific controversy, was to dominate North American vertebrate palaeontology until almost the end of the century, and it has been recounted in much detail in several books (Colbert, 1968; Lanham, 1973; Shor, 1974; Howard, 1975). The very different personalities of the two protagonists certainly contributed much to the bitterness of their rivalry.

Edward Drinker Cope (1840–97) was the son of a wealthy Quaker. He became interested in natural history at a very early age, and as a student at the University of Pennsylvania he attended Joseph Leidy's courses in anatomy. His first scientific work was in herpetology, when he recatalogued the collection of the Academy of Natural Sciences in Philadelphia. In 1863 he was sent to Europe by his father, possibly to avoid military service in the Civil War. There he visited several major museums and perfected his zoological knowledge. On his return to the United States in 1864 he took up farming for some time, but this did not last long, and in 1867 he decided to devote his whole activity to natural history. However, he did not enjoy the teaching position he held for some time at Haverford College, and soon became a scientist of independent means, supported mainly by family wealth. When his father died in 1875, Cope inherited a considerable fortune, but his investments in mining stock proved injudicious and he finally lost almost everything he had. He had to seek a paid position, and in 1889 became professor of geology and mineralogy at the University of Pennsylvania. In 1896, one year before his death, he became professor of zoology and comparative anatomy, thus occupying Leidy's former chair. Cope has been described as a 'fiery, eccentric genius', and there is no doubt that individualism in an extreme form was one of his most salient characteristics. He was a nervous man, who could not concentrate on one subject very long and often made errors because of hasty judgements. He enjoyed naming new species, often on inadequate material, and it is reported that he named more than a thousand species of fossil vertebrates (many of them have turned out to be synonyms). Cope was impatient with administrative duties, which explains some of his problems with various institutions. His aggressive manner readily made him obnoxious to some people. He liked scientific controversies, and probably enjoyed his almost life-long fight against Marsh. His strong religious feelings probably influenced some of his broad generalisations about the process of evolution.

Othniel Charles Marsh (1831–99) also developed an interest in natural history at an early age when he was living with his father and stepmother in western New York. Thanks to financial support from his maternal uncle George Peabody, he was able to study natural sciences at Yale University. In the 1860s he spent almost three years in Europe, visiting museums and purchasing fossils. In 1866 his uncle was persuaded to donate $150,000 to Yale University to establish a natural history museum (now called the Peabody Museum in his honour). At the same time, Marsh became professor of palaeontology. This, however, was not a paid position, and for most of his career Marsh was financially supported by his uncle's fortune.

Marsh's personality was quite different from Cope's. He was a calm person, who worked slowly and carefully. A solitary man, he had few friends, and many visitors found him aloof and overbearing. In contrast to Cope, who did most of his scientific work by himself, Marsh had many assistants, and several of his most famous monographs were partly 'ghost-written' by them (which eventually led to personal problems and denunciations of Marsh's autocratic behaviour toward his employees). Marsh was an ardent and jealous fossil collector, who easily became suspicious of visitors to his collections.

Cope and Marsh had become acquainted with each other during their stay in Europe and for a short time their relations were friendly. They even dedicated species to each other (Shor, 1974). In 1868 Cope, who at that time lived in Haddonfield (where Leidy's *Hadrosaurus* had been found), took Marsh to Cretaceous localities in New Jersey. He later complained that when he returned to the localities afterward, he found that he could no longer obtain fossils there because Marsh was paying for them. This may indicate the beginning of their rivalry, which became worse when in 1870 Marsh noticed that Cope had restored a skeleton of the plesiosaur *Elasmosaurus* from Kansas with the vertebrae reversed and the head at the end of the tail. According to Marsh, Cope's pride was so hurt when the error was pointed out to him that he became Marsh's bitter enemy. Whatever the exact origins of the feud, it developed over the years as both men became interested in the same fossil localities in the West. After Marsh had collected late Cretaceous vertebrates in Kansas in 1870, Cope quickly followed and hired one of Marsh's guides, thus initiating a competition for able field assistants which was to last for many years. In 1872 the feud was renewed when Cope and Marsh both became interested in the Bridger Basin of Wyoming, which was yielding strange, previously unknown Eocene mammals, such as the uintatheres, on which Leidy was working. This marked the end of Leidy's palaeontological career, as he was disgusted by the turn events were taking. In his desire to describe new taxa before Marsh, Cope even resorted to telegraphing his scientific papers, written in the field, to Philadelphia for publication, with the predictable result that some of the new names became strangely garbled, and that synonyms accumulated, which resulted in nomenclatural chaos. Cope's peculiar ideas about scientific publications angered Marsh, who became obsessed with publication dates, and their correspondence shows how mere competition quickly degenerated into outright hostility when the two men accused each other of stealing fossils, antedating their publications, and so on. When remains of large Jurassic dinosaurs were discovered in Colorado in 1877, both Cope and Marsh were notified at about the same time, and this coincidence, of course, fuelled their quarrel. Huge bones had been found near the town of Morrison by a schoolmaster by the name of Arthur Lakes, who sent specimens to both Cope and Marsh, but finally decided in favour of Marsh. Cope had to send Lakes's fossils to Marsh, which infuriated him. However, another schoolmaster, O.W. Lucas, had found even larger bones in the same formation near Canyon City, and these he sent to Cope. Eventually, an even better locality was discovered by two men working for the Union Pacific Railroad, W.E. Carlin and W.H. Reed. The place was called Como Bluff, and there an outcrop of the Morrison Formation extending for 7 miles (11km) was to yield tonnes of well-preserved dinosaur bones (Colbert, 1968). Marsh's assistant Samuel Wendell Williston, who was to become a famous vertebrate palaeontologist, was sent there to investigate. He was enthusiastic, and urged Marsh to reach an agreement with Carlin and Reed, according to which they were to excavate dinosaurs for him at Como Bluff and to keep other collectors away from the locality. Excavations there were continued until 1889, and yielded an enormous amount of excellent material.

Although his collectors spied on the activities of Marsh's assistants, Cope was unable to find an equally good locality in the Morrison Formation, but he did obtain some good dinosaur specimens from the much younger late Cretaceous Judith River Beds of Montana, which he had first visited in 1876. At the time, just after Sitting Bull's victory over Custer at Little Big Horn, the region was supposed to be unsafe because of Indians, but this did not deter Cope, who spent a successful field season there with Charles Sternberg, then at the beginning of his career as a 'fossil hunter'. They found duck-billed dinosaurs, as well as some of the first remains of ceratopians (Colbert, 1968). Marsh also collected dinosaurs in the late Cretaceous formations of the West. Among his major discoveries was that of the horned dinosaur *Triceratops*, even though he at first identified horn cores of this animal as belonging to an extinct species of bison (Colbert, 1968).

The rivalry extended to fields other than Tertiary mammals and Jurassic and Cretaceous dinosaurs. The discovery of Permian vertebrates in New Mexico was the occasion for an outburst of renewed hostility when Marsh claimed in print, after he had heard a lecture by Cope on the subject, that he, Marsh, had been the first to identify them (Shor, 1974).

Finally, the feud between Marsh and Cope, which had hitherto been restricted to scientific circles and to the deserts of the West, became public in 1890 when a newspaperman named William Hosea Ballou started a press campaign against John Wesley Powell (then director of the US Geological Survey) and Marsh (who then worked for the Survey) in the columns of the *New York Herald* under headlines of 'Scientists wage bitter warfare' (a detailed study of this 'fossil feud' in the press has been published by Shor, 1974). The campaign had been initiated by Cope, who publicly accused Powell and Marsh of ignorance, plagiarism and incompetence. The US Geological Survey, with Powell at its head and Marsh working for it as a palaeontologist, had been under attack for some time, with Congress suspecting misuse of public funds and political manipulation. Powell and Marsh answered Cope's accusations in the same vein; Marsh's assistants, who apparently had reasons to resent his treatment of them, were drawn into the controversy, various eminent American scientists were asked for their opinion, and the battle went on for a fortnight in the columns of the *New York Herald*. The general public, however, does not seem to have been very responsive to that kind of scientific brawl, and the controversy soon lost momentum, at least as far as newspaper articles were concerned. Many of the scientists who had unwittingly been drawn into the quarrel found it deplorable that dissensions within their community should make headlines in the popular press. Neither of the protagonists emerged victorious, and both Cope and Marsh had to continue their scientific careers with relatively limited means until their deaths. As remarked by Shor (1974), in the final decade there was not peace, but there was no longer open conflict.

Seen in retrospect, the feud between Cope and Marsh has many ludicrous aspects, but it should not conceal the important contributions they both made to vertebrate palaeontology. The western parts of the United States, with their vast exposures of fossiliferous rocks, provided opportunities which were unequalled at the time for palaeontological research. The scattered and limited outcrops of Europe could not compare with the thick, well-exposed American series, and many of the gaps in the fossil record which had bothered Darwin were filled by the researches of Cope, Marsh and their co-workers.

The American localities provided the first complete skeletons of many groups of fossil vertebrates which had hitherto been known only from fragmentary remains. The need to collect more or less complete specimens, often under difficult circumstances, also led to technical innovations. When excavating dinosaurs on the Judith River in 1876, for instance, Cope and Sternberg realised that it was safer to encase bones, together with some of the surrounding matrix, in some sort of protective 'jacket' before they were shipped to the laboratory. This ensured safer transportation. For this they used cloth strips dipped into rice that had been boiled into a thick paste. Marsh's collectors used plaster of Paris instead of rice paste, and the 'plaster jacket', which is still in use today, was thus invented in the 1870s.

Improved collecting techniques contributed to the spectacular progress of vertebrate palaeontology in North America. An important result of the discovery of abundant mammalian remains in Tertiary deposits was the establishment of a succession of Cenozoic faunas, which could be paralleled with the European sequence which was also beginning to emerge. Cope, especially, devoted a huge monograph ('Cope's Bible', 1884) to the Tertiary vertebrates of the West, and tried to correlate the mammal faunas he discovered with those of the Old World. Among his most spectacular discoveries was that of early Tertiary mammals in the 'Puerco Formation' of New Mexico, which was older than the Eocene; this revealed a whole new chapter (corresponding to the Palaeocene epoch) of mammalian history.

The fairly complete North American record also yielded remarkable series of fossil vertebrates which provided convincing evidence in favour of evolutionary transformations. In 1868 Marsh had had the opportunity to obtain some fossil horse remains, found when digging a well, during a brief stop at a railway station in Nebraska (Howard, 1975), and this had made him interested in horse evolution. After a few years of collecting in various Tertiary basins, he had assembled enough fossils to reconstruct an evolutionary lineage leading from early Tertiary forms to the modern horse. This all-American family tree of the horse contrasted with phylogenies based on European fossils, such as that put forward by Kovalevskii, which led from *Palaeotherium* to *Equus* via *Anchitherium* and *Hipparion*. Huxley had accepted Kovalevskii's phylogenetic reconstruction, but when he visited the United States on a lecture tour in 1876, Marsh's impressive collection of fossil equids convinced him that his views had to be revised. In a lecture on the proofs of evolution, he acknowledged that the stages of equine evolution were better represented in North America than in Europe, and used Marsh's fossil ancestors of the horse as one of the most demonstrative proofs of the reality of evolution: a virtually continuous series of transformations, involving toe-reduction and tooth-complication, could be reconstructed. In that case, evolutionary theory could even be shown to have a predictive value: Huxley postulated that a very early four-toed form of horse would be found in deposits older than the Eocene ones which had yielded *Orohippus*, then the most primitive, three-toed, horse ancestor in Marsh's series. This was confirmed by Marsh's discovery of *Eohippus* in early Eocene deposits (it was only later that it became apparent that *Eohippus* was identical with Owen's *Hyracotherium*, from England). Since then, the phylogeny of the horse has become a classic of evolutionary palaeontology, mentioned in all textbooks — although further research soon showed that the unilinear model leading from *Eohippus* to *Equus* through *Orohippus, Mesohippus, Miohippus, Protohippus* and *Pliohippus* was an oversimplification (Simpson, 1951).

Evidence in favour of evolution could also be found among the newly discovered Mesozoic vertebrates of North America. One of the groups which most interested Huxley in this respect were the toothed birds from the Niobrara Chalk of Kansas. There, besides remains of mosasaurs, giant turtles and huge pterosaurs, Marsh had found abundant remains of marine birds which he described as *Ichthyornis* and *Hesperornis* in a long monograph published in 1880. The toothed birds from Kansas could be added to *Archaeopteryx* as spectacular evidence of the descent of birds from reptiles. In Darwin's words, Marsh's work 'on these old birds and on the many fossil animals of North America had afforded the best support for the theory of evolution, which has appeared within the last 20 years' (quoted by Gregory, 1979).

Although he provided much paleontological evidence in favour of evolution and was a staunch supporter of Darwinism, Marsh did not publish much on the more general or abstract aspects of evolutionary theory. One of his few contributions on this subject was a recognition of a 'law' of brain-size increase in Tertiary mammals. Cope, apparently, was more interested than Marsh in the mechanism of evolution, but possibly because of his religious convictions, he rejected natural selection and turned to a neo-Lamarckian view of evolutionary change. He made great use of recapitulation theory, and envisaged evolution as a succession of additions to the growth of the individual, with use and disuse as a major factor in the acquisition of new characteristics. Moreover, conscious choice was seen as directing evolutionary change (Bowler, 1984). Within this framework there was no place left for chance and random change, and Cope endeavoured to recognise regularities in evolution. One of those he thought he could demonstrate became known as Cope's Law, and enounced that in vertebrate lineages there was usually an increase in size from earlier to later forms. Empiric statements of this kind led to more general conclusions about increasing specialisation and orthogenetic evolution: the vertebrate lineages revealed by fossils from the American West seemed too unidirectional to be explained by the conjunction of random variation and natural selection, and other directing forces were called for. The search for so-called 'laws of evolution' and orthogenetic series was to become a frequent feature of palaeontological research at the end of the nineteenth century and the beginning of the twentieth.

To the general public, however, the most spectacular results of Marsh's and Cope's researches was the discovery of whole new extinct faunas, which were often represented by remarkably preserved specimens. The 'extinct monsters' from the American West could be reconstructed with more certainty than many of their previously known European counterparts, which were often known from rather scanty remains. This was especially true of the abundant dinosaurs whose enormous skeletons had been excavated from the Jurassic and Cretaceous strata of the West. Marsh's last big monograph (1896) was devoted to them, and these discoveries were the subject of much enthusiastic comment in popular and semi-popular books on palaeontology on both sides of the Atlantic. However, very little of the huge collections brought together by Cope and Marsh became available for public display in their lifetimes (Gregory, 1979).

In the long run, the bitter rivalry between Cope and Marsh probably did more good than harm to the progress of vertebrate palaeontology. As remarked by Lanham (1973), the 'durable, intelligent, well-focused hatred' between the two men was 'a long-range creative force as powerful as love' which led to many important discoveries. It must be deemed fortunate, however, that the fighting ended with the deaths of the protagonists, and that the next generation of American palaeontologists agreed to share the immense fossil resources of the United States in a more peaceful way. Many of the emerging young palaetontologists of the 1890s had worked for either Marsh or Cope, or had been their students, but the feud between the old masters did not extend to their successors.

The development of large palaeontological museums (or palaeontological sections in natural history museums) was an important characteristic of the last decade of the nineteenth century and the opening one of the twentieth century. Cope's collections were purchased by the American Museum of Natural History in New York in 1895, and Marsh's collections went partly to the Peabody Museum at Yale, and partly (for what he had collected when employed by the US Geological Survey) to the US National Museum in Washington, DC. Soon, other natural history museums in the United States, such as the Carnegie Museum in Pittsburgh and the Field Museum in Chicago, started to acquire extensive collections of fossil vertebrates. The public impact of mounted skeletons of dinosaurs and other large extinct animals was (and still is) great, and collecting went on unabated in the rich fossiliferous regions of the West. Among the scientists who were instrumental in bringing about this new trend in vertebrate palaeontology, Henry Fairfield Osborn (1857–1935) occupies a prominent position. As an undergraduate at Princeton in 1877, he had organised with his friend W.B. Scott a palaeontological expedition to the Bridger Basin, and this was the beginning for both of them of long careers in vertebrate palaeontology. Both Osborn and Scott were disciples of Cope, who had befriended them when they still were students. After the then usual stay in European universities, both returned to Princeton, where Scott remained for the rest of his life and became a renowned expert on fossil mammals. In 1891 Osborn joined the staff of the American Museum of Natural History, where he founded the Department of Vertebrate Paleontology. Among the men who joined the department at that time were several who contributed greatly to the development of American vertebrate palaeontology both in the field and in the laboratory: the names of William Diller Matthew, Walter Granger and Barnum Brown are closely associated with this phase of the development of the American Museum, during which much successful collecting was done both in the United States and in foreign countries. Among the early spectacular field discoveries by the American Museum team was that of the Jurassic dinosaur locality which became known as Bone Cabin Quarry, in Wyoming. The locality was found in 1898 and was exploited until 1905. The abundant material collected there allowed a spectacular group of Jurassic dinosaur skeletons to be mounted at the American Museum of Natural History

(Colbert, 1968).

Osborn's influence on North American vertebrate palaeontology in the first decades of the twentieth century was considerable. He published a large number of papers on fossil reptiles and mammals, including extensive monographs on such groups as titanotheres and proboscideans. Beyond purely descriptive work, he also published on more general topics, and contributed to the development of the tritubercular theory of mammalian teeth (which had been initiated by Cope) and to the establishment (with Matthew) of a succession of Cenozoic land mammal ages for North America. Osborn's contribution to evolutionary theory has been largely forgotten; although he was influenced by Cope's ideas and attributed an important role to orthogenesis, he cannot be called a Lamarckian. He thought that both Lamarck's and Darwin's speculations on the origin of new characteristics were 'flatly contradicted' by palaeontological evidence (Osborn, 1930). The mechanisms he envisaged (involving what he called 'rectigradations' and 'adaptive allometrons') were not especially clear. His evolutionary conceptions were put forward at a time when genetics was still in its infancy, and the ulterior development of the synthetic theory of evolution quickly made them largely obsolete (this applies to many similar 'theories' of evolution put forward by palaeontologists at the beginning of the twentieth century which were often characterised by much confusion between pattern and process). What remains of Osborn's general ideas is mainly the concept of adaptive radiation.

The considerable research work done at the American Museum of Natural History was emulated by other large natural history museums which had often been founded thanks to the generosity of private patrons. The museum founded by Andrew Carnegie in Pittsburgh was especially active in the field of vertebrate palaeontology, and one of its great successes was the discovery and exploitation of an extraordinary dinosaur locality in the Morrison Formation of Utah, which was first known as the 'Carnegie Quarry' before becoming the Dinosaur National Monument. The site was discovered by Earl Douglass, who was on the staff of the Carnegie Museum and excavated it for several years, with the result that several excellent dinosaur skeletons were collected.

Among the most spectacular dinosaur skeletons in the

Carnegie Museum was a composite *Diplodocus* specimen from the Jurassic of Wyoming. Andrew Carnegie decided to have several plaster casts made of it, which were donated to various natural history museums all over the world (Desmond, 1975). Casts of *Diplodocus carnegiei*, as the animal was called, were thus mounted, under the supervision of the palaeontologist W.J. Holland in London, Paris, Berlin, Vienna, Bologna, La Plata and Mexico City. This in itself is revealing of the status vertebrate palaeontology had gained in natural history museums: large skeletons of extinct animals, especially dinosaurs, had become first-class exhibits, which drew large crowds of visitors.

The popularisation of vertebrate palaeontology was made easier in the United States not only by the availability of remarkably preserved spectacular specimens, but also by a conscious effort to reconstruct the creatures of the past in an attractive way. Ever since Cuvier's first attempts at producing life restorations of the Montmartre ungulates, vertebrate palaeontologists had tried to provide reconstructions of the extinct animals they studied, but these had too often been based on too scanty material, or were rather amateurish, not to say crude, from an artistic point of view. At the end of the nineteenth century joint work between capable artists and palaeontologists resulted in reconstructions which were both aesthetically pleasing and scientifically more acceptable. This was not an exclusively American development, as testified for instance by the interesting reconstructions prepared in England by J. Smit for the popular books on fossil vertebrates by the Reverend Hutchinson (1894, 1897). However, the paintings produced in the United States by Charles Knight in the 1890s and the first decades of the twentieth century, were of such quality that they quickly became classics of palaeontological reconstruction. Knight's large paintings and murals adorned many natural history museums in the United States, and they were abundantly reproduced in popular articles and books. The result was that the supposed appearance in life of many extinct reptiles and mammals became familiar to a wide public. As early as 1914, a sauropod dinosaur even appeared on the cinema screen in Winsor McCay's early animated cartoon *Gertie the Dinosaur*. Ever since 'extinct monsters' had first been reconstructed at the beginning of the nineteenth century, they had strongly excited the imagination of laymen as well as scientists, but interest in them had long been restricted to the more educated members of

139

society who had some scientific background. With the development of palaeontological museums and the spread of reconstructions of fossil vertebrates, these became part of the cultural background of larger circles who knew little about science in general, but were impressed by the sheer size and bizarre shapes of dinosaurs and other large extinct animals. This popularisation, of course, was accompanied by some unavoidable distortion. Although this phenomenon was especially marked there, it was not restricted to the United States. In Europe, during the later part of the nineteenth century and the beginning of the twentieth, both important new discoveries and the growth of public palaeontological collections also led to an increase in the public interest in fossil vertebrates.

8

European Vertebrate Palaeontology Before the First World War

Public natural history museums had become established in many European cities in the first decades of the nineteenth century, and had known various fortunes. Among the miscellaneous collections which were displayed to the public, fossil vertebrates, when available, were considered as good exhibits, provided that they were complete enough to be easily intelligible to the layman. When the growth of collections led to the enlargement of the museums, and often to the erection of spacious new buildings to house them, the palaeontology displays usually occupied much space and played an important part in the new museums. When the natural history collections of the British Museum were moved to South Kensington in the early 1880s, Richard Owen, as superintendent of the new British Museum (Natural History), had plans for an 'index museum', which would have illustrated 'the principal modifications of the great Vertebrate sub-kingdom' (Stearn, 1981). Because of opposition from the keepers of the various departments, these plans were not carried out. The palaeontology collections were the responsibility of the Department of Geology, and palaeontology soon became the major preoccupation of the staff of this department (this fact was finally reflected by a change of designation from Geology to Palaeontology in 1956). The importance of the vertebrate palaeontology collections was illustrated by the catalogues which were published by several experts during the 1880s and 1890s: the four-volume *Catalogue of the Fossil Reptilia and Amphibia*, published by Richard Lydekker between 1888 and 1890, for instance, was nearly 1,200 pages long. Fossil vertebrates, being both bulky and spectacular, occupied a prominent position in the exhibition galleries, and their popularity was demonstrated by many contemporary cartoons (Stearn, 1981). Huge mounted skeletons of fossil vertebrates were fast becoming the popular symbol of natural history museums, in Britain as elsewhere.

At the Muséum d'Histoire Naturelle in Paris, fossil vertebrates were at first part of the comparative anatomy collections (Cuvier had held the Comparative Anatomy chair). When a chair of palaeontology was created in 1853, Alcide d'Orbigny, the first professor, had no collection to administer. He built up a large collection of fossil invertebrates, but when, after d'Archiac and Lartet, Gaudry became professor of palaeontology in d'Orbigny's chair, the palaeontology department still had no collection of fossil vertebrates, and the specimens he had collected at Pikermi were turned over to the comparative anatomy department under Paul Gervais. When Gervais died in 1879, Gaudry finally obtained the supervision of the fossil vertebrates and devoted much effort to the establishment of a palaeontology gallery. At first, a provisional gallery was installed in an old building in 1885, but Gaudry had grander plans, and in 1898 a large new building was inaugurated to house the palaeontology and comparative anatomy galleries, as well as the palaeontology laboratories and research collections. Gaudry's museological conception was a direct reflection of his philosophical ideas on evolution: the vast, undivided palaeontology gallery was to give the visitor a vivid impression of the progress of life through geological time. The rapid increase of the collections, with the arrival of such large specimens as Carnegie's *Diplodocus* cast, soon resulted in overcrowding: in 1910, Thevenin already remarked that the building was 'much too exiguous'.

Similar large palaeontological sections appeared in most major natural history museums in Europe in the later part of the nineteenth century. Germany alone had several such museums, with especially large vertebrate palaeontology collections in Stuttgart (with remarkable collections of late Triassic and early Jurassic vertebrates), Frankfurt, Berlin and Munich. With the development of large museums of that kind, ambitious acquisition programmes were often started, and fossil vertebrates from distant places were usually added to the local collections. To mention only one instance, Wellnhofer (1980) reports that while Karl Alfred Zittel was director of the Bavarian State Collection from 1866 to 1904, this institution obtained fossil mammals from the Tertiary and Pleistocene of Uruguay and Argentina, Tertiary mammals from the Dakotas and the Greek island of Samos, Permian amphibians and reptiles from Texas, plesiosaurs from the Jurassic of England, and various Cretaceous and Tertiary vertebrates from Egypt (in addition to Bavarian fossils). The conquest of vast colonial empires in Africa and Asia of course facilitated the development of such 'international' collections, as will be described in Chapter 9. However, spectacular discoveries were still being made in Europe, and even local museums could acquire important collections of vertebrate fossils, provided that they were located in a suitable region. In Normandy, for instance, the museums at Caen and Le Havre were able to display large numbers of fossil fishes and reptiles from the Jurassic strata of the region (Buffetaut, 1983). Similarly, the vertebrate palaeontologist H.E. Sauvage built up a large collection of local fossil vertebrates in the museum at Boulogne sur Mer. In Europe, however, outcrop conditions could not compare with those of the American West or of the desert regions of other parts of the world, but the considerable development of some branches of economic activity partly compensated for the difficulties of fossil collecting in the densely populated European countries. The exploitation of several kinds of mineral resources led to the chance discovery of many important fossil localities, and the comparatively primitive mechanisation of quarries and mines still allowed a relatively easy recovery of well-preserved specimens.

The need for a more efficient agriculture, able to sustain rapid population growth, resulted in a diligent search for natural fertilisers, and the phosphate-rich rocks which were actively exploited sometimes also turned out to be rich sources of vertebrate fossils. A good example from England is provided by the Cambridge Greensand, a Cenomanian formation which, as its name implies, is found in the vicinity of Cambridge and contains at its base abundant dark nodules composed of calcium phosphate. These nodules had erroneously been called 'coprolites', a term which became current even in commercial parlance, and during one period of the nineteenth century they were actively exploited in large pits as a source of fertiliser. After the overburden had been removed, the nodules were dug out and washed, and fairly abundant fossils could be collected *in situ* or recovered from the heaps of washed material (Cowper Reed, 1897). Besides various invertebrates, the 'coprolite bed' usually

143

yielded fragmentary remains of various fossil vertebrates, some reworked from older deposits, others contemporaneous with the deposition of the Greensand. When he was teaching at the University of Cambridge in the 1860s, Harry Govier Seeley was able to obtain thousands of vertebrate bones from the workmen, who were willing to preserve the fossils whenever they were 'adequately remunerated', to use Seeley's term (Seeley, 1901, p. 176). Among the specimens thus acquired were remains of fishes, dinosaurs, plesiosaurs, ichthyosaurs, turtles, crocodilians and the pterosaurs which were Seeley's favourite group. Seeley thought that he had thus assembled for the Sedgwick Museum 'the finest collection of remains of these animals in Europe' (Seeley, 1901, p. 176), and there is no doubt that the three-dimensional preservation of the bones from the Cambridge Greensand greatly added to the knowledge of pterosaurs, whose bones had hitherto mostly been found in a more or less crushed condition on slabs of lithographic limestone. Unfortunately, the heyday of 'coprolite' exploitation in the Cambridge region did not last very long. In 1897 Cowper Reed remarked (p. 122) that

> the importation of phosphate of lime from America and Belgium and the small demand for the material owing to agricultural depression have caused many of the pits to be closed, and the majority of the pits mentioned in the Survey Memoir are now deserted or filled up.

With the closure of the pits, the source of fossil vertebrates, of course, dried up.

The story of the exploitation of the early Tertiary phosphorites of the Quercy region of south-western France is in some respects similar to that of the Cambridge Greensand vertebrate finds. The phosphorites of Quercy are found in 'pockets' in the Jurassic limestones of this dry and stony plateau. Phosphorite was first found there in 1865 near the small town of Caylux, and commercial exploitation began in several places in 1870. About forty pockets were thus excavated (Piveteau, 1951). Although there was much controversy about the formation of the phosphorite in the 1870s (some geologists accepted an organic origin whereas others developed strange theories involving the eruption of phosphate-rich mineral waters), it soon became recognised that the abundant vertebrate remains found in some of the pockets were those of animals which had fallen and died in karstic fissures or caves, or had been transported there by running water. The exploitation of the phosphorite pockets as a source of fertiliser led to the discovery of enormously abundant remains of amphibians, reptiles, birds and mammals. Although articulated skeletons were comparatively rare, the bones were usually well preserved, and in some instances even the shape of soft parts was preserved as casts in phosphate. The rich mammalian fauna from Quercy, with its marsupials, creodonts, carnivores, ungulates, bats, rodents and primates, filled some palaeontologists with enthusiasm. In the introduction to his monograph on these mammals, which was published in 1876, H. Filhol exclaimed (p. 43) that

the localities of Quercy must be considered as having yielded the most interesting evidence hitherto discovered in Europe for the study of fossil mammals, and the animal forms which they reveal are no less valuable than those which have been brought to light in America in recent years.

One of the main problems, however, was that it quickly became apparent that the mammals of the phosphorites were not all of the same age: some were late Eocene in age, others were Oligocene. As the fossils were obtained from the workmen and little systematic field work was done by palaeontologists, the exact provenance of many specimens was not known, and a biostratigraphic zonation of the phosphorite pockets was not possible. After 1890 commercial exploitation of the Quercy phosphorites sharply declined and finally ceased, because of the competition of cheaper phosphates from other regions (notably the North African phosphates, which were also an important source of fossil vertebrates). It was not until the 1960s that the phosphorite pockets of Quercy were exploited again, this time for purely scientific purposes when palaeontologists from the universities of Montpellier and Paris started systematic excavations there, which resulted in a much better understanding of the chronological succession of the Quercy faunas.

This pattern of intense collecting (more often than not by workmen rather than by palaeontologists) in the wake of commercial exploitation, followed by more or less rapid decline when the economic viability of the fossil-bearing rocks went down, was repeated in many instances. The rise and fall of the palaeontological productivity of a famous fossil locality as a consequence of commercial vicissitudes has been studied in some detail in the case of the lithographic limestone quarry at Cerin, in the southern Jura mountains of France (Bourseau, Buffetaut, Barale, Bernier, Gaillard, Gall and Wenz, 1984; Barale, Bernier, Bourseau, Buffetaut, Gaillard, Gall and Wenz, 1985). Late Jurassic fossil fishes were found there as early as 1838, and in the 1850s the quarry was exploited intensively to obtain lithographic stones. When the German palaeontologist Albert Oppel visited the locality in 1854, he saw many quarrymen at work there, and noted that the lithographic stone was sold under the deceptive name of 'Munich stone', on account of its resemblance to the famous lithographic limestones of the Altmühl valley in Bavaria. Like their Bavarian counterparts, the Cerin limestones yielded abundant and beautifully preserved remains of plants, fishes and reptiles. Fossils were bought from the quarrymen or the quarry owners by scientists from Lyons (the nearest large city with an important natural history museum) and other places. An extensive collection of fossil vertebrates from Cerin was thus built up at the Lyons natural history museum by Victor Thiollière and his successor Claude Jourdan. Contracts were made between the quarry operators and the museum, and for 200 francs a year most of the fossils found at Cerin were sent to Lyons. Over the years, however, many fossils from Cerin found their way into palaeontological collections in other places (Andrew Carnegie, for instance, bought an important private collection of fossil fishes from Cerin for the Pittsburgh Museum in 1903). In 1875 the lithographic limestone quarry at Cerin was still prosperous, but financial difficulties quickly led to a sharp decline in the activity of the exploiting companies, and in 1892 the palaeontologist Lortet complained that 'unfortunate circumstances' had severely restricted the exploitation of the quarries, with the result that their palaeontological production had dropped considerably. Among the unfortunate circumstances may have been the rise of photographic methods of reproduction, which resulted in the decline of lithography at the end of the nineteenth century. Be that as it may, the Cerin quarry was abandoned for many years, until the University of Lyons reopened it in the 1970s to start systematic palaeoecological investigations.

The most spectacular of all palaeontological discoveries in late nineteenth-century Europe was also a direct consequence of the exploitation of mineral resources: on 1 April 1878 miners working at 322 metres (1,056 feet) below the surface in a coalmine at Bernissart (Belgium) came across a clay-filled pocket in which they found what they interpreted as gold-filled tree trunks. These eventually turned out to be large fossil bones with pyrite inside; one of the discoverers, miner Jules Créteur, found that they looked like 'ribs more rounded than those of our oxen'. Some fragments were brought to a local café, where they were examined by officials of the mine who were convinced that they were indeed fossil bones. Scientific authorities were notified, and in May 1878 the palaeontologist P.J. Van Beneden concluded that the dinosaur *Iguanodon* was present in the Wealden pocket at Bernissart (Buffetaut and Wouters, 1978; Decrouez, 1981). The director of the Belgian Royal Museum of Natural History, E. Dupont, sent his collaborator L. de Pauw to Bernissart to supervise the extraction of the dinosaur bones with the assistance of the administration of the mine. The fossils occurred in Wealden clays which filled natural pits in the Carboniferous rocks, and digging them up in the mine posed obvious problems which were made even worse by flooding. After their position had been carefully recorded, the skeletons were divided into large blocks containing bones and the surrounding matrix, which were then encased in plaster and circled with iron bands. The blocks were then taken out of the mine and transported to Brussels, where they were prepared. Work was stopped in 1881, because no space was available to store more fossils, and the renewed excavations which had been planned were never carried out (Norman, 1985). Preparation of the fossils was made especially difficult by the presence of pyrite, which decomposed when it was exposed to the air, so that an elaborate treatment had to be devised to prevent decay (this ultimately proved insufficient, and in the 1930s the fossils had to be treated again to ensure their preservation). Altogether, twenty-nine more or less complete skeletons of *Iguanodon* were recovered, and mounting them for display also proved a difficult task, as much space was needed to work on the large specimens. The first mounted skeleton was displayed in 1883, and a new wing was added to the natural history museum in 1902 to accommodate a group of mounted specimens and skeletons shown in the position in which they had been found. Besides the *Iguanodon* skeletons, many other fossil vertebrate remains were found at Bernissart: a theropod dinosaur, 5 crocodilians, 5 turtles, one amphibian and 3,000 fishes. The Bernissart discovery attracted much attention in scientific circles and from the public. The Paris Museum expressed great interest in these remarkable fossils which had been found so close to the French border, and negotiations were started by Gaudry to obtain a skeleton for the Paris collection. This, however, was prevented by a vigorous press campaign in Belgian newspapers, and like many other institutions, the Paris Museum had to be content with a plaster cast. The reptiles from Bernissart, and especially the *Iguanodon* specimens, were described by the palaeontologist Louis Dollo in a series of short notes. The Bernissart excavations permitted considerable progress in the knowledge of *Iguanodon* and dinosaurs in general. *Iguanodon* had hitherto been known by rather incomplete specimens, and the various reconstructions of this animal by Mantell and Owen have been mentioned in Chapter 5. The complete specimens from Bernissart allowed much more accurate reconstructions, and the bipedal silhouette of this dinosaur soon replaced the older quadrupedal restorations in palaeontology textbooks (nevertheless, the recent researches of D.B. Norman, 1980, on the Bernissart specimens suggest that *Iguanodon bernissartensis* spent much of its time in a quadrupedal position).

The discovery of spectacular concentrations of fossil vertebrates in the course of the commercial exploitation of economically valuable deposits remained a relatively unusual occurrence, however. In many cases, finds of good vertebrate specimens were relatively rare events, and constant supervision had to be maintained on the work done in mines or quarries so that the fossils could be recovered at the right time. In such circumstances, professional palaeontologists could hardly be expected to keep permanent watch over individual localities, and the role of amateurs was of paramount importance. This is clearly exemplified by the story of the Leeds collection (or rather collections) of fossil reptiles from the Jurassic Oxford Clay of the Peterborough area in England (Leeds, 1956). There, the Oxford Clay was (and still is) quarried in huge open pits and used for brick-making. From time to time, the workmen chanced upon more or less complete remains of reptiles such as ichthyosaurs, plesiosaurs or mesosuchian crocodiles. During the last decades of the nineteenth century, several professional and non-professional palaeontologists became involved in the collection and study of these fossils, but the most important discoveries were made by two brothers, Charles and Alfred Leeds, who had no special training in palaeontology: Charles had been a lawyer before he finally emigrated to New Zealand where he engaged in farming, and Alfred had taken over the family farm at Eyebury. The Leeds brothers lived close to the vast clay pits which, at the end of the nineteenth century, were exploited with crowbars, a method far less destructive than the steam-diggers which began to be introduced at the beginning of the twentieth century. The pits were regularly inspected for fossils by Alfred Leeds, and workmen sometimes notified him of important finds by mail or telegram. The fossils were cleaned and mended at Eyebury, and over the years considerable numbers of excellent specimens of various Jurassic reptiles were thus accumulated. The Leeds collection soon became known to some of the leading palaeon-tologists of the time, and Harry Govier Seeley, Arthur Smith Woodward, Henry Woodward and J.W. Hulke were frequent visitors to Alfred Leeds's house. Eventually, a first large collec-tion, weighing five tonnes, was purchased by the British Museum in 1890. The marine reptiles in this collection were described in a two-volume monograph by C.W. Andrews (1910–13). Alfred Leeds went on collecting for more than twenty years after the sale of this first collection. Officials from the Department of Geology of the British Museum visited him at least once a year, and the specimens they did not want (usually duplicates of forms already in the first collection) could be sold to other buyers. A fossil dealer from Bonn, G. Stürtz, thus acquired many good skeletons, which were then prepared and sold to various institutions; for this reason, Oxford Clay vertebrates can now be seen in Tübingen, Vienna, Paris and elsewhere. After Alfred Leed's death in 1917, a second collection was offered for sale, and specimens were purchased by several British institutions.

Alfred Leeds never wrote any scientific descriptions of the fossil vertebrates he discovered. Other amateur palaeontologists of the period, however, studied and described their finds, which were sometimes of great importance. Thus, the interesting late Palaeocene vertebrates from the Rheims region were discovered and studied by a local physician, Victor Lemoine, in the 1870s, 1880s and 1890s. However, the growth of an extensive palaeon-tological literature and the increasing specialisation of the discipline was making it more and more difficult for most amateurs to keep abreast of the rapid developments in the study of fossil vertebrates. Even among the academics, the 'general practitioners' of palaeontology were disappearing slowly, to be replaced by specialists.

The growth of large palaeontological museums led to an increased demand for good specimens which could serve as spectacular exhibits, and the commercial value of such fossils consequently became considerable. In areas where good material could be obtained easily and abundantly enough, fossil localities even became exploited on a commercial basis. The most famous instance of this is the exploitation of the late Liassic *Posidonienschiefer* of the Holzmaden region, in south-western Germany. As mentioned above, well preserved vertebrate skeletons had been discovered in these bituminous black shales as early as the eighteenth century, and it was soon appreciated that the fossils from the quarries around Holz-maden, Boll, Ohmden and other villages of the area were of both remarkable scientific interest and great aesthetic value. The bituminous shale was distilled to produce oil, and the more resistant layers were quarried to make slabs which in the past had been used as roofing slates, but increasingly were being used for interior decoration. In the 1890s the son of a quarry owner, Bernhard Hauff (1866–1950), after receiving some training in palaeontological preparation in Stuttgart under Oskar Fraas, started to collect and prepare fossils in a systematic way and to sell them to museums and universities (Hauff, 1953; Schneider-Hauff, 1973). Spectacular fossils from Holzmaden, on their slabs of dark grey shale, are now to be seen in most important palaeontological institutions, and a remarkable museum has been built by the Hauff family in Holzmaden itself. It is estimated that more than three hundred ichthyosaur skeletons were thus prepared in Bernhard Hauff's workshop. Although ichthyosaurs are the most abundant reptiles in the Holzmaden shales, skeletons of crocodilians, plesiosaurs and pterosaurs were also unearthed, not to mention fishes and invertebrates. Despite some excessive reconstructions on incomplete remains, the specimens prepared by Bernhard Hauff and his associates were an important source of scientific information, as shown by the number of scientific papers devoted to them by such leading German palaeontologists as Eberhard Fraas, Friedrich von Huene and Fritz Berckhemer. Among the major discoveries made on Holzmaden fossils collected by Hauff was that of the outline of the skin of ichthyosaurs, which demonstrated that these reptiles possessed dorsal and caudal fins.

At the end of the nineteenth century, intense collecting activity was not limited to the western European countries where vertebrate palaeontology had developed. Fossil vertebrate localities were discovered and systematically excavated in parts of eastern Europe where comparatively little work had hitherto been done. Some of the discoveries were entirely due to chance: in 1895, for instance, fossil bones were found by a young Hungarian lady, Ilona Nopcsa, on the family estate at Szentpéterfalva in the Hatszeg valley of Transylvania (a region which at the time belonged to the Austro-Hungarian Empire). Ilona's brother Ferenc (or Franz — he used the Hungarian and German version interchangeably), then aged 18, searched 'frantically' for more remains, and soon found many interesting specimens, including a good skull. When he returned to school in Vienna in the autumn, he took the bones with him and showed them to the famous geologist Eduard Suess (Tasnadi Kubacska, 1945). Suess was enthusiastic about the specimens: they were remains of late Cretaceous dinosaurs. The Viennese Academy of Sciences decided to send the palaeontologist Gustav von Arthaber to investigate the locality and study the fossils. However, no funds were forthcoming, as the Academy apparently hoped that Arthaber would travel to Transylvania at his own expense. The plans for a scientific investigation came to nothing, which keenly disappointed young Nopcsa, and he resolved to visit Suess again. The latter's suggestion was simple: Nopcsa should describe his discoveries himself. The problem was that young Franz had no knowledge whatsoever of palaeontology or osteology. Undaunted, he immediately decided to study vertebrate palaeontology, and within one year his first scientific paper (a description of the skull of a hadrosaur from Szentpéterfalva, published in 1899) was ready. This marked the beginning of Baron Nopcsa's brilliant scientific career (Weishampel and Reif, 1984), which was made somewhat erratic by his activities as a secret agent for the Austro-Hungarian Empire (Tasnadi Kubacksa, 1945). The Szentpéterfalva finds were spectacular enough: in Maastrichtian clays and sandstones, Nopcsa found concentrations of reptile bones (which he attributed to the predatory activity of crocodiles) containing remains of hadrosaurs, iguanodontids, sauropods and armoured dinosaurs, as well as turtles, crocodiles and birds. This was the best find of late Cretaceous dinosaur fauna in Europe, and during the first decades of the twentieth century Nopcsa published a number of papers on it, although political events (the annexation of Transylvania by Romania after the First World War) prevented him from exploiting it fully. Political, professional and personal problems eventually put an unbearable strain on Nopcsa's highly strung personality, and he committed suicide after murdering his secretary in Vienna in 1933.

In contrast to the discovery of the Transylvanian dinosaurs, Amalitsky's finds of Permian reptiles in northern Russia were the result of carefully planned investigations. Vladimir Prochorovitch Amalitsky (1860–1917) was professor of geology in Warsaw and in 1895 started a programme of geological and palaeontological research in the late Palaeozoic deposits of northern Russia, between Moscow and Arkhangelsk. He was especially interested in continental deposits, in which earlier explorers of this part of Russia, such as Murchison or Keyserling, had found no fossils. Amalitsky succeeded in finding abundant remains of terrestrial and freshwater organisms. Among them were some stegocephalian amphibians and a variety of reptiles which strangely resembled some of the forms previously described from the South African Karroo. The existence of reptiles in the Permian of Russia had already been reported by several palaeontologists, including Owen and Seeley, but the pareiasaurs, dicynodonts and large carnivorous therapsids found by Amalitsky were both numerous and well preserved. He recognised that they had important palaeobiogeographical implications, as they suggested that terrestrial connections had existed between the southern continents (Gondwanaland) and northern Russia in the Permian. In 1899 he obtained 500 rubles from the Imperial Society of Naturalists in St Petersburg, and an equal sum from the Ministry of Public Instruction, to start systematic palaeontological excavations (Amalitsky, 1900). These took place on the banks of the Lesser northern Dvina, near the town of Kotlas. There, a steep escarpment exposed a vast sandstone lens in which animal and plant remains of Permian age could be found. During visits between 1896 and 1899, Amalitsky had discovered numerous bones and plant fossils in sandstone concretions which had fallen from the escarpment down to the towpath along the river. In the summer of 1899, in the company of his wife, he travelled to Kotlas and started excavations with the assistance of a number of workmen. The work was quite dangerous, as sandstone blocks kept falling from the steep bluff. It was therefore decided to excavate the lens from the top. When a trench was dug there, a further difficulty was encountered: at a depth of 1.5 metres (5 feet), the ground was permanently frozen and ice filled all the cracks and cavities. After removing frozen rocks down to a considerable depth, they finally found a layer of nodules. The first of them were barren, but soon the skeleton of a large pareiasaur was found in several big concretions. This was the first of a spectacular series of finds. Careful excavation revealed that many of the skeletons were oriented in the same direction, a fact that Amalitsky interpreted as showing that the animals had been buried together in the bed of a river. When the excavations stopped in August 1899, thirty-nine groups of bones had been recovered, and the whole collection, when it was shipped to Warsaw by train, weighed 20 tonnes. Preparation of the bones proved a difficult, lengthy and dangerous task (Woodward, 1918):

> With much trouble Professor Amalitsky engaged and trained some skilled masons to extricate the fossil skeletons from the intractable matrix in which they were embedded, and more than one unfortunately succumbed to the effect of the peculiar siliceous dust which the work produced.

The most abundant reptiles found on the Dvina were pareiasaurs. Some were relatively small, but very large forms, 4 or 5 metres (13 or 16 feet) long, were also found. Carnivorous therapsids (gorgonopsians) were well represented, some of them reaching a length of 3 metres (10 feet). Dicynodonts also occurred, as well as some amphibians. Preparation took so long that when Amalitsky died in 1917 he had been able to describe only a small part of the collection, and descriptive work had to be continued by other Russian palaeontologists in the 1920s.

Amalitsky's excavations on the Dvina River were not the only large-scale palaeontological expeditions conducted in the Russian Empire at the turn of the century. In August 1900 a native Siberian found the frozen carcass of a mammoth on the banks of the Beresovka River, in north-eastern Siberia. In 1901 the Academy of Sciences in St Petersburg sent an expedition, under Herz and Pfizenmayer, to excavate this remarkable specimen, which was brought back entire to St Petersburg (Pfizenmayer, 1926).

The considerable field activity of European vertebrate palaeontologists at the end of the nineteenth century and the beginning of the twentieth resulted in an enormous amount of descriptive work. Specialised palaeontological journals had become established in the major European countries and provided a convenient outlet for that kind of research. Lavishly illustrated monographs were published on various faunas or zoological groups. The number of recorded species of fossil vertebrates increased tremendously, and knowledge of many groups which had hitherto been represented by scanty remains increased considerably. This often led to important systematic revisions. It became apparent, for instance, that the large Mesozoic terrestrial reptiles which Owen had lumped into a single group, the Dinosauria, actually showed considerable diversity. The dinosaurs obviously had to be split into several groups, and in the 1870s and early 1880s Huxley, Marsh and Cope all put forward new classifications (Colbert, 1968). It was Seeley, however, who in 1887 recognised that the dinosaurs could easily be divided into two major groups, the Saurischia and the Ornithischia, on the basis of the structure of the pelvis, and this classification is still used today.

The striking new results of vertebrate palaeontology were summarised in two volumes of Zittel's influential *Handbuch der Palaeontologie*, published in 1890 and 1893. Zittel's work was soon translated into English and French. By comparison with earlier works of the same kind, such as Pictet's *Traité de Paléontologie*, the second edition of which had appeared in 1853, quantitative and conceptual progress was obvious.

One of the major reasons why the abundance of new forms could be ordered in a more satisfactory way, of course, was that the concept of evolution had become generally accepted among palaeontologists. The need for repeated affirmation of the reality of evolution based on palaeontological evidence grew less urgent as more and more intermediate forms and evolutionary lineages were discovered. Evolution was taken for granted as the implicit framework within which palaeontologists worked. The availability of more abundant fossil material, which filled many previous gaps in the record, led to the realisation that many of the earlier phylogenies had been hastily reconstructed, and represented successions of general evolutionary stages rather than true genealogical lineages. The case of the evolution of the horse, first reconstructed by Kovalevskii on the basis of almost purely European material, and then re-interpreted in the light of Marsh's North American finds, has already been mentioned.

These re-interpretations sometimes led to conflict between the old generation of palaeontologists, who had given the first evolutionary interpretations of the fossil record, and their younger colleagues who tried to use a stricter methodology. This is clearly shown, for instance, by the controversy which developed in 1905 at the Société Géologique de France between Albert Gaudry and Charles Depéret. Depéret (1854–1929) was professor at the University of Lyons and had become famous for his work on Tertiary mammals, partly based on the excavations he had organised in several localities in southern France. In 1904 he had described new material of the Eocene perissodactyl *Chasmotherium* from one of these localities, and reviewed the evolution of tapir-like mammals. This led him to remark (1904, pp. 579–580) that Gaudry had

> intercalated an Eocene animal of the lophiodontid group as forming a link between a Miocene tapir and a Pliocene tapir. No example serves better to warn against the theoretical conceptions (often invoked by Mr Gaudry in his works) which consist in determining the age of a fossil by examining its *evolutionary stage*.

This method, Depéret claimed, could sometimes be useful when carefully applied to forms truly belonging to a well-attested lineage. More often than not, however, it was used on fossils which belonged to separate lineages, in which evolution had not proceeded at the same pace, if not in different directions, and this resulted in serious errors.

Gaudry felt that Depéret's criticism had been unjust and expressed in unkind terms, and in 1905 he reasserted his belief in the usefulness of his method and regretted that a disciple should try to diminish the achievements of his 'old master'. In his answer, Depéret (1905) deplored that the controversy should be brought by Gaudry on to an emotional ground where he could not follow him, and reasserted that he did not think that Gaudry's methods were 'favourable to the progress of palaeontology'. In his book *Les transformations du monde animal* (The transformations of the animal world), published in 1908,

Depéret reviewed previous works on evolutionary palaeontology. Concerning Gaudry, he admitted that his works, because of a certain 'poetical charm', had been influential in turning the opinion of French palaeontologists toward evolution, but he pointed out that many of his generalisations on progressive evolution were imbued with an 'exaggerated and sometimes even somewhat naïve sentimentalism'. Gaudry's phylogenetic reconstructions had often been 'artificial filiations' linking morphological stages taken from different lineages. On the whole, Depéret thought that Gaudry's ideas on evolutionary palaeontology had been much inferior to Cope's. Following Zittel, Depéret urged palaeontologists to be more careful in reconstructing filiations which too often were based on the hasty examination of insufficient evidence. The importance of convergent evolution, in particular, had often been underestimated.

As recognised by Depéret, the enthusiastic acceptance of evolution by many palaeontologists had too often led them to premature conclusions. The development of life through geological time had been far more complex than was supposed by some broad generalisations. The concept of a regular progression, for instance, was contradicted by many new discoveries which showed that evolution had been a branching process, in the course of which transformations had followed various directions and taken place at different rates. The need nevertheless was felt to search for general laws, which would bring some order into the increasing confusion. Cope had already thought that laws of size-increase and growing specialisation could be deduced from the fossil record. Many other palaeontologists of the late nineteenth and early twentieth centuries similarly tried to discover such laws. Among them was Louis Dollo (1857–1931), who described the *Iguanodon* from Bernissart. Dollo had been born in France, where he became a mining engineer. He then moved to Belgium, and became 'assistant naturalist' at the Royal Museum of Natural History in Brussels in 1882. In 1886 he became a Belgian citizen. Dollo's work was mainly on the fossil vertebrates from Belgium. His training as an engineer had given him a taste for a very methodical, almost mathematical, exposition of his researches, and he had no doubt that evolution was determined by law. In 1893 he clearly expressed his famous law of the irreversibility of evolution: once a lineage had become engaged in a specialised direction, it could under no circumstances 'go back' to a less specialised condition. There were simple and spectacular examples of this irreversibility: the horse, having lost its side toes, could not possibly regain them; sirenians could not grow again their hind limbs which had become vestigial in the course of evolution, etc. Irreversible evolution sometimes resulted in fairly complicated processes: as shown by Dollo (1901), in some sea turtles the dermal armour had been greatly reduced in a first stage, but in later forms of the same group a secondary development of bony plates, not homologous with the original ones, had taken place.

The optimistic belief that generally valid laws of evolution could be deduced from a careful study of the fossil record was widespread among palaeontologists. A consequence of it was that orthogenetic models of directional evolution and the hypothesis of an unknown inner force directing the evolution of groups and determining their duration were easily accepted. The idea that phyletic lineages went through a regular succession of phases (their 'career' as Depéret termed it in 1908) became popular, and this resulted in the rather anthropomorphic distinction of periods of youth, maturity and senility in the evolution of zoological or botanical groups. This superficial analogy between individual development and evolution at the specific or supraspecific level provided an easy explanation for the phenomenon of extinction, which had been a matter of controversy since the eighteenth century: just as an organism, after growing old, eventually dies, so a species (or a higher group) first became senile and then died out. Specialisation was seen as an inherently nefarious process, which almost inevitably led to extinction. The huge antlers of the Irish elk, the large size and often bizarre appearance of many dinosaurs, the bony protuberances of the skull of uintatheres, to mention but a few examples, were interpreted as hyperspecialised features which were forerunners of inevitable extinction. Although most palaeontologists now reject such views, they sometimes still find an echo in popular palaeontological literature.

This kind of fairly rigidly determined evolution, guided by mysterious laws, was hardly compatible with the Darwinian processes which had been accepted by at least some of the first evolutionary palaeontologists. As noted by Depéret in 1908, most naturalists preferred the Lamarckian theories, and palaeontologists, moreover, stressed the importance of unknown internal forces in addition to direct environmental factors. This must be considered within the general context of the 'eclipse of Darwinism', to use Julian Huxley's expression, which characterised evolutionary thought at the end of the nineteenth and the beginning of the twentieth centuries (Bowler, 1984). Disaffection for Darwinian hypotheses was widespread among neontologists, and palaeontologists, who had little to say about the actual *mechanism* of evolution, also felt that its course, as revealed by fossils, was easier to explain in neo-Lamarckian terms. Random variation and natural selection were deemed insufficient to account for the kind of directional evolution which orthogenetic lineages implied, and outstanding vertebrate palaeontologists such as Depéret or Dollo expressed ideas which were explicitly or implicitly neo-Lamarckian. This situation was to last until the emergence of the synthetic, or neo-Darwinian, theory of evolution in the 1930s and 1940s. In retrospect, the contribution of vertebrate palaeontology to evolutionary theory between 1880 and 1920 may be considered as relatively small. The reconstruction of an ever increasing number of phyletic lineages among various groups did provide some kind of confirmation of evolutionary change through geological time, whether the reconstructions were correct or not, but the mechanism of change was beyond the reach of palaeontologists; the 'laws' they discovered were at most empirical rules, and exceptions to them were numerous. Too strict adhesion to these so-called 'laws' easily led to erroneous conclusions. The dwarf elephants of the Pleistocene of some Mediterranean islands, for instance, had been interpreted, notably by the German palaeontologist Pohlig, as 'degenerate' races of the normal-sized elephant species of the continent; isolation on islands supposedly was responsible for size-reduction (although the mechanism through which insular endemicity caused dwarfism was — and to some extent still is — a matter of discussion). However, the British palaeontologist Dorothea Bate had suggested that the 'dwarf' elephants were actually survivors of a small ancestral stock which had given rise to the large elephants of the continent, and that they had survived on the islands because of the protection afforded by isolation. This unconvincing hypothesis was enthusiastically accepted by Depéret (1908), because it was in agreement with the so-called 'law' of size-increase postulated by Cope: if Bate's conception was correct, there was no need to call for a 'law of size decrease', which other palaeontological evidence did not suggest. The desire to explain the observed pattern of evolution by the operation of a small number of fairly strict laws thus led to a loss of flexibility in palaeontological interpretation, and the number of facts which contradicted the so-called laws was bound to increase as new fossils were found.

Another field which vertebrate palaeontologists began to explore more fully around the turn of the century was that of the reconstruction of the mode of life of extinct animals. This concern was as old as palaeontology, as shown by eighteenth-century speculations on the diet of the mastodon, but early speculations had often been based on insufficient analysis — hence the rather fanciful restorations which adorned many nineteenth-century books. Good examples of such 'artists' conceptions' of extinct creatures are to be found, for instance, in the popular French books on palaeontology by Louis Figuier (1866) and Camille Flammarion (1886): seen through the eyes of imaginative artists, dinosaurs and other 'prehistoric monsters' easily took on an appearance more reminiscent of mythological dragons than of living creatures. As mentioned above, the discovery of better-preserved specimens (notably in the American West) led to more reliable reconstructions, and an American school of palaeontological reconstruction emerged at the end of the nineteenth century. In Europe too some vertebrate palaeontologists became conscious of the need for more accurate reconstructions of the appearance and mode of life of extinct vertebrates. Foremost among them was Louis Dollo, who endeavoured to develop a special branch of palaeontology which he called 'ethological palaeontology' (his follower the Austrian Othenio Abel preferred the term 'palaeobiology', in a rather different meaning from that in which it is used today). The primary aim of Dollo's researches was not the preparation of graphic restorations for the general public, but the reconstruction of the adaptations of extinct beings. This, as Abel (1912, 1917) showed, could provide useful data for phylogenetic reconstruction. The basis of Dollo's and Abel's work was functional anatomy, and the search for living analogues of extinct animals was an important part of the process. Dollo applied his methodology to many of the fossil vertebrates he studied, including mosasaurs from the Belgian Cretaceous deposits and the Bernissart *Iguanodon*. A famous anecdote illustrates the way in which Dollo used his morphological analogies between fossil and living vertebrates to reconstruct the way of life of the latter. During a conversation with Leopold II, the king told him that he believed *Iguanodon* was a sort of giraffe. Not unexpectedly, Dollo's answer (Dollo, 1923, p. 76) was diplomatic:

Yes, sire, but *reptilian giraffes*, because they were scaly animals, as reptiles usually are — not hairy beasts, as mammals ordinarily are. But they did find their food in the foliage of trees, as giraffes do — although by other means.

In Dollo's opinion, the king's interpretation of the 'palaeoethology' of *Iguanodon* was fundamentally correct. Expanding on a suggestion originally put forward by Mantell, Dollo thought that *Iguanodon* had possessed a long prehensile tongue similar to that of the giraffe (an opinion not supported by recent research: see Norman, 1980).

Despite the unavoidable mistakes in a field where guesswork necessarily plays such an important part, the works of Dollo and Abel in the field of 'palaeobiology' were important. They drew attention to the fact that studies of the anatomy of fossil vertebrates could be more than a basis for systematic work and phylogenetic reconstruction; if conducted within the proper functional scope they could lead to more reliable reconstructions of the way of life of extinct animals. Modern functional palaeoanatomy essentially is a refined version of Dollo's 'ethological palaeontology'.

The study of the past distribution of animals is another branch of palaeontology which gained a strong new impetus around 1900. As mentioned earlier, Buffon had already discussed possible dispersal routes for extinct vertebrates, and Cuvier had recognised that migrations could have played an important part in faunal replacement. The geographical distribution of fossil vertebrates had been an important element in the palaeontological evidence used by Darwin in the *Origin of Species*. Nevertheless, and for obvious reasons, a very large part of the early work on fossil vertebrates had been done on European material. With the discovery of remarkable fossil localities in America, the situation had changed, and Europe gradually had lost the central position it had seemed to occupy for lack of data from other continents. Vertebrate evolution obviously had to be envisaged on a worldwide basis, and it became apparent that many problems posed by European (or for that matter North American) fossils might have their solution elsewhere. Such problems were further complicated by the fact that geography could no longer be considered as a stable, unchanging background to organic evolution. Although very few authors were ready at the time to envisage continental displacements (for exceptions, see Muir Wood, 1985), it had long been understood that the relative positions of land and sea had changed considerably throughout geological time. Palaeontology was one of the main sources of evidence for constructing palaeogeographical maps such as those published by the Austrian palaeontologist Melchior Neumayr in 1895. Such maps showed that during some periods of the geological past continental areas which today are widely separated by oceanic basins had been connected by land bridges. Conversely, such past connections between the continents necessarily had had a major influence on the geographical distribution of terrestrial vertebrates. A more complete knowledge of the zoogeographical history of vertebrate groups was an obvious prerequisite to a better understanding of their evolution. Palaeontological discoveries in South America, Africa, Asia and Australia were putting the European and North American fossil records in their proper perspectives — and at the same time were raising new problems. A thorough exploration of these 'new worlds' was needed, and the worldwide expansion of palaeontological research was also an important feature of the development of this science in the later part of the nineteenth century.

9

Vertebrate Palaeontology in the Age of Imperialism

As we have seen in Chapter 7, North America was the first non-European area where vertebrate palaeontology developed, and North American contributions to the new science soon became of great importance. Although the Luján *Megatherium* described by Cuvier had played no negligible part in the early history of vertebrate palaeontology, the beginnings of this science in South America were rather slow. After the former Spanish and Portuguese colonies had gained their independence in the first decades of the nineteenth century, it took quite a long time for local palaeontologists to appear. Much of the palaeontological exploration in the first half of the century was thus carried out by Europeans. As early as 1817–20, the German explorers Johann Baptist von Spix and Carl Friedrich Philipp von Martius, who had been sent on an expedition to Brazil by the King of Bavaria, discovered the famous early Cretaceous localities of the Ceara region in the north-eastern part of the country, which since then have yielded thousands of fossil fishes, as well as interesting reptiles (Spix and Martius, 1828). The palaeontological researches of Charles Darwin in Argentina have already been mentioned. Another important European discoverer of South American fossil vertebrates was the Dane Peter Wilhelm Lund (1801–80), who has been hailed as the 'father of Brazilian palaeontology' (Hoch, 1984). Lund had studied medicine and physiology in Copenhagen, after which he had gone to Brazil for three years, during which time he investigated the zoology and botany of the region around Rio de Janeiro. Back in Europe, he visited Cuvier in Paris in 1830, and was influenced by his ideas on catastrophism (Hoch, 1984). When he returned to Brazil in 1833, Lund soon became engaged in palaeontological excavations.

With the botanist Riedel, he set out on a prolonged expedition to the interior of Minas Gerais. There, in 1835 they met another Dane, Peter Clausen, who had settled there as a landowner, and who told them about fossiliferous caves in the vicinity of the town of Lagoa Santa. Bones had been found there by the local inhabitants who used the cave soil as fertiliser, and this had given rise to the usual tales of giants. Clausen himself collected the fossils to sell them to museums in Europe. Lund settled in Lagoa Santa and started to work on the abundant late Pleistocene and early Holocene fossils from the caves. He thus brought together an extensive collection which was later sent to Denmark. An abundant and varied assemblage was unearthed, comprising both extinct and still-living forms. Among the extinct animals were mastodons, ground sloths and a South American horse. Even before it was fully described, the Lagoa Santa fauna attracted much attention, as it was one of the first fossil mammal assemblages to be found in a tropical country. Darwin, in particular, was impressed by the relations between the Lagoa Santa mammals and the living animals of South America. Moreover, human remains were found in association with some of the animal bones, and they played some part in the controversy about fossil man which raged during the middle part of the nineteenth century (although recent research seems to indicate that the Lagoa Santa men are actually Holocene in age). Lund did not publish much on his Brazilian discoveries; complete descriptions of his finds were given after his death by Danish zoologists (Simpson, 1984). Lund could never bring himself to leave Brazil again, and he died there in 1880.

Other European expatriates also played an important part in the early development of South American vertebrate palaeontology. European naturalists settled in Argentina too and greatly contributed to the palaeontological exploration of the country. One of them was the Frenchman Auguste Bravard, who had worked on Tertiary mammals from central France, but was exiled because of his opposition to Louis-Napoléon's *coup d'état* in 1851. Bravard collected fossil mammals in the Pampaean formation of Argentina, but he was prevented from publishing full descriptions of his finds by his untimely death in the Mendoza earthquake of 1861. Bravard's material eventually came into the hands of another European expatriate, Hermann (or German) Burmeister (Biraben, 1968). Burmeister (1807–92) was born in Prussia, and spent the first forty years of his life there. He studied natural history and medicine in Greifswald and Halle, became briefly interested in politics in 1848, soon grew disillusioned, and embarked on a first South American journey in 1850. This first contact with South America was ill-fated, as Burmeister broke his right leg in an accident in the vicinity of Lagoa Santa in 1851; this left him lame for the rest of his life. A second journey, from 1856 to 1860, took him to the 'States of La Plata', as Argentina was then called, through which he travelled widely. On his return to Halle, where he held a university position, he was soon faced with serious administrative problems, which led him to resign in 1861 and to settle permanently in Argentina. There, he became director of the Public Museum of Buenos Aires, which had been founded in 1812 and was to become the National Museum. Burmeister devoted himself to the organisation of the museum, with the intention of turning it into a major natural history institution. He published a large number of papers on a variety of subjects, ranging from geography to ornithology and entomology. His palaeontological researches had been started in Germany, where he had worked on trilobites, labyrinthodonts and Jurassic crocodilians. In Argentina he became interested in the large mammals from the Pleistocene of the pampas, and published a number of papers on them. Although he was not born in Argentina, he was one of the first naturalists to describe the fossil vertebrates of this country 'on the spot', so to speak; previous descriptions had been based on material sent to Europe by European explorers.

This important trend in South American vertebrate palaeontology was continued by two major figures of Argentinian natural history at the turn of the century, Francisco P. Moreno (1852–1919) and Florentino Ameghino (1854–1911). They had very different careers and eventually became bitter enemies, but in different ways they both greatly contributed to the discovery of the palaeontological riches of their country. Moreno was a field man and an organiser, who in the 1870s explored remote parts of Patagonia at a time when government troops were still fighting the Indians (in 1880, he was even taken prisoner by an Indian chieftain and barely escaped with his life). During his travels in Patagonia he heard curious native tales which apparently referred to extinct mammals. The mythical Elengassen, for instance, was described as a large and dangerous monster covered with heavy armadillo-like armour; Moreno thought that the story was based on distant reminiscences of glyptodons

(Moreno, 1979). He was keenly interested in the natural history and anthropology of his country, and was the founder and first director of the Museo de La Plata, a magnificent natural history museum comparable to those which were being built at the same time in Europe and North America. The museum opened in 1884, and under Moreno's active supervision it quickly acquired extensive vertebrate palaeontology collections. Pleistocene and late Tertiary mammals made up the bulk of the collection, but older vertebrates were also collected. In 1889, for instance, two collectors employed by the museum, Steinfeld and Botello, were sent to the region of the Rio Chubut in Patagonia to collect dinosaur bones which had been reported by Carlos Ameghino, Florentino's brother. In Moreno's words (1891b, p. 61):

> To transport the bones of the enormous dinosaurs discovered on a previous trip, animals whose size is measured in tens of metres, and whose bones cannot be carried on horseback because of their great weight, a 12-ton flat-bottomed boat and a small but strong cart, appropriate to these regions, were built in our workshops.

Boat and cart were transported in several pieces to the foothills of the Andes and reassembled there. Steinfeld and Botello then proceeded down the Rio Chubut, collecting fossils on the way. One of the best specimens they found was the arm of a large sauropod, 3.2 metres (10ft) high.

Although Moreno did not publish much on vertebrate palaeontology, his activity as head of the La Plata Museum was of considerable importance in establishing a centre for palaeontological research which is still very active today.

Florentino Ameghino (Ingenieros, 1957; Simpson, 1984) was a very different character from Moreno. He was the son of Italian immigrants, and was born a few months after his parents had arrived in Argentina and settled in Luján (where Cuvier's *Megatherium* had been found). He was self-taught, and started to collect fossils at an early age. His first palaeontological publication, in the 1870s, drew sharp criticism from Burmeister. In 1878 Ameghino attended the Universal Exhibition in Paris, where he met several leading French scientists. He had brought with him an important collection of Pleistocene mammals, and sold several of them to Cope, who had also come to Paris for the Exhibition. Back in Argentina in 1882, Ameghino opened a bookshop (with the unusual name 'El Gliptodon') to earn a living, but his main activity was vertebrate palaeontology, and he published a large number of papers and books on the subject. In 1884 he became professor of natural history at the University of Cordoba, but left his position in 1886 to become vice-director and head of the palaeontology department of Moreno's La Plata Museum. This state of affairs did not last long, however, as he soon quarrelled with Moreno. The latter claimed that Ameghino had sold the museum a collection of fossils but refused to produce the catalogue and that he tried to use the museum collectors for his own purposes (Moreno, 1891b). Be that as it may, Ameghino resigned in 1887, and opened another bookshop in La Plata. In 1902 he finally became director of the National Museum in Buenos Aires, and stayed there until his death.

Florentino Ameghino's work was largely a result of his collaboration with his brother Carlos (1865–1936). Carlos Ameghino was a skilled fossil collector and was employed in that capacity by the La Plata Museum when his brother became vice-director there. In 1887 he went on his first expedition to Patagonia. This trip was remarkably successful and Carlos Ameghino was able to collect a number of Tertiary mammals definitely older than any previously found in South America (they belonged to the Miocene Santa Cruz fauna). After he resigned from the La Plata Museum together with his brother, Carlos Ameghino went on collecting fossil vertebrates in Patagonia at the family's expense until 1903. He thus discovered a tremendous number of Tertiary vertebrates, which were described by his brother and revealed a succession of faunas completely new to science (Simpson, 1984). Florentino Ameghino did not usually accompany his brother in the field, but nevertheless did so in 1903 when he joined Carlos's last expedition to Patagonia.

Florentino Ameghino was a convinced Darwinist — he went so far as to consider Darwinism as an 'exact science' — but his phylogenetic reconstructions were influenced by his desire to give South America, and more particularly Patagonia, a central place in mammalian evolution. In his 1906 book on the sedimentary formations of Patagonia, driven by what Simpson (1984, p. 92) has called 'an unfortunate, unbreakable obsession', he proposed that many groups of Old World and North American mammals, including horses, elephants and carnivores, had their origins in Patagonia. Man was no exception: as early as the 1870s Ameghino was convinced that he had found evidence of the great antiquity of the human genus in Argentina, and he finally reconstructed a wholly South American human phylogeny, based on various specimens which he thought were of Tertiary and Pleistocene age, and to which he gave names such as *Tetraprothomo* and *Diprothomo* (Ingenieros, 1957). Some of these so-called prehumans were in fact relatively recent human bones, while others were based on misidentified animal remains. Ameghino's eccentric phylogenetic and biogeographical conceptions were linked with equally peculiar ideas concerning the age of the fossil-bearing formations of Patagonia. Although his studies on Tertiary faunal succession in southern South America were a major contribution to the unravelling of vertebrate history in that part of the world, he consistently over-estimated the antiquity of the faunas his brother Carlos and others were discovering. He insisted, for instance, that his *Notostylops* and *Pyrotherium* beds were late Cretaceous in age, whereas modern palaeontologists place them in the Eocene and Oligocene, respectively (Simpson, 1984). Claims that Tertiary mammals had been contemporaneous with dinosaurs in Patagonia were partly based on the misidentification of dinosaur-like crocodilian teeth, but Ameghino's main motivation was probably his unconscious desire to make Patagonia the main centre of mammalian evolution. Despite numerous errors and misinterpretations, however, his scientific work was of considerable importance in the development of South American vertebrate palaeontology.

At the time, however, some other palaeontologists obviously thought that Ameghino's undue multiplication of new taxa and his sometimes highly disputable interpretations were rather a hindrance to the progress of scientific knowledge. Thus, when Richard Lydekker was invited by Moreno to study fossil vertebrates in the La Plata Museum in 1893 and 1894, he strongly disagreed with Ameghino's conclusions on South American ungulates, and said so in the preface to his monograph on the subject in the following terms (Lydekker, 1893, p. VII):

> Since my work speaks for itself, it is unnecessary to refer to it here, but I may mention that it serves to show how cautious palaeontologists must be in accepting genera and species described by Argentine writers as valid.

This was clearly aimed at Ameghino, and Moreno must have enjoyed it, but Lydekker's harsh judgment has caused resentment among many Argentine palaeontologists (Bondesio, 1977).

Despite the rapid growth of a local school of vertebrate palaeontology, foreigners were also active in Argentina about the turn of the century. British palaeontologists such as Lydekker and Smith Woodward studied fossil mammals and reptiles at the La Plata Museum. Foreign institutions also sent collectors to the rich Patagonian localities. The Frenchman André Tournouër, for instance, collected Tertiary mammals there for the Paris Museum (they were the last fossils Gaudry worked on just before his death in 1908). This period also saw the beginning of a series of American palaeontological expeditions to Patagonia: in the late 1890s John Bell Hatcher, who had previously worked for Marsh in the American West, collected fossils there for Princeton University, and the important material he found formed the basis of a series of monographs edited (and partly written) by W.B. Scott (Simpson, 1984).

Argentina was in some respects exceptional among South American countries because of the early development of large local palaeontological institutions. In other countries this took longer, and at the beginning of the twentieth century much of the research work in vertebrate palaeontology was still done by foreign specialists. In Brazil, for instance, interesting fossil vertebrates were collected in the Cretaceous basins along the Atlantic coast by Joseph Mawson, a British railway engineer, one of the many working throughout South America on the railways then being built by British firms. Mawson sent his finds (including remains of coelacanths and giant crocodilians) to Arthur Smith Woodward, who studied and described them in London.

In the colonial empires of the main European powers, the growth of vertebrate palaeontology followed various paths, depending on the importance of the European population and the general development of the colony, as well as on its palaeontological resources.

Early explorers and travellers sometimes had the opportunity to collect fossil vertebrates. Well before Burma fell under British control, for instance, remains of late Tertiary mammals had been found there by John Crawfurd, who had been sent on an embassy to the Burmese capital of Ava in 1826–7. The British envoy had been able to collect 'no less than seven large chests full of fossil wood and fossil bones' along the course of the Irrawady as he went up that river in a steam-boat from Rangoon to Ava (Buckland, 1828). The fossils were sent to the Geological Society of London, where Buckland and Clift examined them. The bones were found to be those of mammals (mastodons, rhinoceroses and deer) and reptiles (crocodiles and turtles). Buckland was much interested in these remains, which to him indicated the former occurrence in Asia of animals very similar to the Tertiary forms from Europe, despite the fact that Europe and Burma now have very different climates.

Not unexpectedly, however, the most spectacular palaeontological discoveries in the British Asiatic colonies were made in India. Although a few vertebrate remains had been found in north-eastern Bengal in the 1820s (Pentland, 1828), the major finds which really attracted the attention of the scientific world, were made in the Siwalik Hills of north-western India in the 1830s. The discovery of the late Tertiary vertebrate localities of the Siwaliks was the result of co-operation between a naturalist, Hugh Falconer, and several army officers, among whom the most active was Captain (later Sir) Proby T. Cautley. Falconer (1808–65) had studied natural history in Aberdeen and Edinburgh, where he obtained the degree of MD. In 1830 he went to India as an assistant-surgeon for the East India Company, and in 1831 became director of the Botanical Gardens at Suharunpoor, 25 miles (40 km) from the Siwalik Hills. He soon embarked on a geological study of the Siwaliks, and came to the conclusion that their detritic deposits were probably of Tertiary age. At first, however, no fossils could be found, but the mention of giant's bones from this area in an old Persian book suggested that remains of large vertebrates might well be discovered. This was verified when Falconer, following Cautley's indications, found bones of crocodiles and turtles. In 1834 he discovered a turtle shell and (Murchison, 1868, p. XXVII) 'immediately after the search was followed up with characteristic energy by Capt. Cautley in the Kalowala Pass by means of blasting, and resulted in the discovery of more perfect remains, including Miocene mammalian genera'.

Soon, other officers stationed in the area joined in the search, and Lieutenants Baker and Durand were able to report the discovery of a great fossil deposit; they had found it after a local Rajah had presented Lieutenant Baker with proboscidean teeth he thought were the remains of giants destroyed by a mythical hero. When Falconer visited the locality, he collected more than three hundred bones in six hours, and grew understandably enthusiastic about it. Thanks to the labours of Falconer and his military friends, a rich vertebrate fauna was thus unearthed from the Siwalik deposits. Reptiles were represented, among others, by crocodiles and gavials, and by an enormous tortoise which received the impressive name of *Colossochelys atlas*; the mammal remains included bones of various proboscideans, rhinoceroses, hippopotamuses, pigs, horses, ruminants (including the huge giraffe *Sivatherium*), carnivores and some of the first fossil monkeys to be discovered. This was the first important Tertiary fauna from a tropical country to be reported, and the papers by Falconer and Cautley, prepared in India under considerable difficulties because of the lack of literature and comparative material, were acclaimed as major contributions to vertebrate palaeontology. Falconer continued his natural history researches in India until he had to leave for Britain on sick leave in 1842. He brought with him five tonnes of fossil bones, but both the British Museum and the Geological Society declined to accept the collection because of lack of space, and most of it was presented to India House, only a few of the most remarkable specimens going to the British Museum. During most of his stay in England Falconer worked on his Indian fossils, but when he had to return to India in 1848 he had not been able to complete the description of all of them. In 1855 he retired from the Indian Service and returned to Britain, where he resumed his palaeontological researches (Murchison, 1868).

Work on the fossil vertebrates of India was continued after Falconer's return to Britain by palaeontologists working with the Geological Survey of India, which published a series of high-quality monographs under the title *Palaeontologia Indica*. Among them was Richard Lydekker, who described not only Siwalik mammals but also older vertebrates such as Cretaceous dinosaurs and the interesting fauna (comprising amphibians, phytosaurs and rhynchosaurs) from the Triassic of central India.

Far from India, the palaeontological exploration of other territories under British rule also proceeded at a rapid pace. In Canada (Lambe, 1905), vertebrate footprints were discovered as early as 1841 in Carboniferous rocks in Nova Scotia by Sir William Logan. They were important in demonstrating that amphibians had been in existence as early as the Carboniferous.

170

Nova Scotia was to play an important part in the discovery of Carboniferous tetrapods: in 1850 William Dawson found the osseous remains of a Palaeozoic land vertebrate in a coalmine at Pictou. This was the beginning of a long series of important finds of Carboniferous amphibians and reptiles, mainly in the famous Joggins locality, where the tetrapod skeletons were preserved inside fossilised tree trunks. Eastern Canada also provided abundant remains of still-older vertebrates: the Devonian agnathans and fishes from Scaumenac (or Escuminac) Bay, on the Gaspé Peninsula (Québec), and from the Campbellton area in New Brunswick, were studied by the Canadian J.F. Whiteaves in the 1880s and 1890s, as well as by British palaeoichthyologists such as R.H. Traquair and A.S. Woodward.

Soon, important discoveries were being made outside the more densely populated regions of eastern Canada, and even in the Arctic regions. During his expedition of 1852–4 in search of John Franklin (who had disappeared in the Arctic in 1847), Captain Sir E.B. Belcher was able to obtain vertebrae and ribs of a Triassic ichthyosaur. The remains were found when building a cairn on the summit of Exmouth Island, at latitude 77°16′N. They were taken to England and identified by Owen (Belcher, 1855). Some two decades later, Andrew Leith Adams (1877) in Dublin described a cervical vertebra of a reptile discovered on Bathurst Island (70°36′N) by Captain Sherrard Osborn as *Arctosaurus osborni*; Lydekker later attributed it to a Triassic dinosaur.

However, the most spectacular discoveries of Canadian fossil vertebrates were made in the western part of the country, when dinosaur remains came to light in the deep valleys of some of the rivers which cut through the prairies of Alberta (Colbert, 1968). The first dinosaur bones discovered in western Canada were found by George M. Dawson, William Dawson's son, who despite being a hunchback was an active field geologist working for the Canadian Boundary Survey and later for the Geological Survey of Canada. In the 1870s he found dinosaur remains in Saskatchewan and Alberta and sent them to Cope for identification. Cope recognised hadrosaur remains. In 1884 Joseph Burr Tyrrell, Dawson's assistant, found fossil bones in the valley of the Red Deer River in Alberta. Among them was the skull of a carnivorous dinosaur, which after an arduous journey was sent to Cope for description. A few years later, in 1888, another geologist of the Canadian Survey, Thomas C. Weston, tried to collect fossils along the Red Deer River valley by floating down the stream on a boat and exploring the outcrops on the banks. The boat sank after a few miles, however, and it was not until the following year that Weston was able to collect an abundance of dinosaur bones with the help of an experienced boatman. In the 1890s more dinosaur remains were collected in the Red Deer River area by Lawrence Lambe, also a member of the Geological Survey of Canada. Lambe's rather fragmentary fossils were described in collaboration with Henry Fairfield Osborn. It was then realised that the so-called 'Belly River' fauna of Canada was older than the Lance Formation of Montana, already explored by Cope. The obvious scientific interest of the dinosaurs from Alberta caused what Colbert (1968) has called the 'Canadian dinosaur rush'. In 1910 Barnum Brown of the American Museum of Natural History, set out on his first collecting trip down the Red Deer River (he too used a specially built boat to explore the outcrops). He was extremely successful, and the expedition was repeated in 1911. The Geological Survey of Canada then became worried about too many dinosaur skeletons from Alberta being sent to New York, and the already famous American collector Charles H. Sternberg was hired to search for dinosaurs along the Red Deer River. From 1912 to 1915 both Barnum Brown and Charles Sternberg collected remarkable dinosaur skeletons in the same area. Despite the competition between the two collecting parties, relations remained friendly, and there was no repetition of the 'fossil war' between Cope and Marsh. Charles Sternberg's three sons went on collecting dinosaurs in Alberta for several years after Brown's expeditions had ceased. Many of the magnificent skeletons of late Cretaceous dinosaurs now exhibited in New York, Toronto and Ottawa were the result of this 'dinosaur rush' from 1910 onwards.

In Australia the search for fossil vertebrates began in the 1830s, when British colonists started to explore the Wellington Caves of New South Wales (Archer and Hand, 1984) and found Pleistocene fossil bones. Some of these were sent to England, where Richard Owen studied them. His first description of Australian fossil mammals was published in 1838, and for many years he worked on Australian fossil vertebrates (1838b). At his instance, money was made available for palaeontological researches by the Colonial Secretary of New South Wales. The discoveries in the Wellington Caves and other localities soon showed that in the

geological past Australia had already been inhabited by marsupials, a fact Darwin was able to use in support of his 'law of
succession of types'. Some of these extinct marsupials were very
large and rather different from existing forms, which sometimes
made them difficult to identify. Thus, in 1843 Owen described a
jaw fragment from the Wellington Caves as that of a primitive
proboscidean allied to *Deinotherium*. A year later, however, he
was proved wrong by the discovery of more complete remains
which showed that the animal (which had been called
Diprotodon) was actually a gigantic marsupial (Hutchinson,
1894). Owen grew rather enthusiastic about this new discovery,
as did others; it is said that F.W.L. Leichhardt, the Prussian
explorer who died when attempting to cross the Australian
continent in 1848, had hoped to discover living representatives of
Diprotodon.

One of the first 'local' vertebrate palaeontologists in Australia
was Gerhard Krefft (1830–81), an emigrant from Germany who
arrived in Australia during the 1852 gold rush and became
curator of the Australian Museum in Sydney in 1861. Krefft
collected fossil mammals in the Wellington Caves, and described
some of them, although most of the material was sent to Owen.
Eventually, however, their co-operation ceased when they
disagreed on the diet of *Thylacoleo*, the strange 'marsupial lion'
from the Australian Pleistocene (Owen, probably rightly,
thought that it was carnivorous, whereas Krefft thought that it
was a plant-eater).

Krefft was one of the first of a series of late nineteenth-century
Australian vertebrate palaeontologists, among whom the most
prominent were De Vis in Queensland, Etheridge in New South
Wales and McCoy in Victoria. They built up important collections, which comprised a few Mesozoic reptiles as well as many
Pleistocene marsupials. An especially important discovery took
place in 1893, when an Aboriginal stockman found large bones at
Lake Callabonna (then called Lake Mulligan), a vast salt-pan in
South Australia (Tedford, 1984). The South Australian Museum
sent the geologist Henry Hunt to investigate, and when he
reported that bones were extremely abundant at Lake Mulligan,
a collecting party was sent there under his leadership. Within
three months, the collectors had counted 360 skeletons of
Diprotodon on a limited area of the lake bed. Remains of
kangaroos and giant flightless birds were also found. The animals
had apparently become mired in the sticky mud of a Pleistocene

lake bed and died on the spot. The director of the South Australian Museum, E.C. Stirling, and his assistant A.H.C. Zietz, were so interested that they went to Lake Mulligan themselves in the autumn of 1893 to continue the excavations. One of the results was that at last the complete skeleton of *Diprotodon* (including the feet, which hitherto had been unknown) could be described. The importance of the locality was recognised by the South Australian government, and in 1901 Lake Callabonna was officially set aside as a scientific reserve.

The giant extinct birds of New Zealand were discovered by European settlers at about the same time as the fossil marsupials of Australia (Mantell, 1851). In the 1830s a missionary, the Reverend W. Colenso, heard strange tales told by the Maoris about a gigantic bird, the moa, which their ancestors had hunted; some moas supposedly were still alive. Colenso and his colleague W. Williams eventually managed to obtain some moa bones, and Colenso published an account of them in 1842. Even before that, however, in 1839, Owen had exhibited at the Zoological Society of London an incomplete femur of an enormous bird which had been brought to him by a Mr Rule. Owen concluded that it had belonged to a flightless bird of the 'struthious' kind, which was more heavily built than the ostrich. These fossils indicated that the extinct fauna of New Zealand had been even more peculiar than the present one, and in the 1840s more remains were collected by several travellers and colonists. In 1846 and 1847 a systematic search for moa remains was conducted by Walter Mantell, Gideon Mantell's son, who had emigrated to New Zealand a few years before. Walter Mantell at first thought that the moa might still be living, but he soon gave up the idea and concentrated on obtaining fossil bones: apparently the moas had been exerminated by the Maori within a few centuries after they had settled in New Zealand. Mantell managed to bring together a collection of between 700 and 800 bones belonging to various species and genera. Some localities proved extremely rich. In a peaty deposit on the seashore north of Otago, for instance, well-preserved, sometimes articulated bones of large moas were found in some abundance, until natives and whalers became 'excited by the large rewards injudiciously given by casual visitors' for the fossils, and started to collect them carelessly and indiscriminately, so that the locality was ruined (Mantell, 1851). Moa bones also could be found in volcanic tuffs; Walter Mantell found such a deposit on the western shore of the North Island, near the mouth of the Waingongoro River. The bones were well preserved but extremely fragile, and great care had to be taken in excavating them. In a letter to his father, Walter Mantell described the difficulties of fossil collecting at this particular locality (Mantell, 1851, p. 102):

Unfortunately the natives soon caught sight of my operations, and came down in swarms — men, women and children — trampling on the bones I had carefully extracted and laid out to dry, and seizing upon every morsel exposed by the spade. My patience was tried to the utmost, and to avoid blows, I was obliged to retreat and leave them in the possession of the field; and to work they went in right earnest, and quickly made sad havoc. No sooner was a bone perceived than a dozen natives pounced upon it, and began scratching away the sand, and smashed the specimen at once. It was with great trouble, and by watching the opportunity of working in the absence of the Maoris, that I procured anything worth having.

In some places moa bones were associated with charcoal and stone tools, which demonstrated that the ancestors of the Maoris had indeed hunted the giant birds.

Remains of much older vertebrates were also found in New Zealand in the nineteenth century (Welles and Gregg, 1971). In 1859 reptile bones were found in the bed of Waipara River in the South Island by Thomas Hood, a Scottish settler, who took them to Sydney, where the local geological authorities declared them to belong to a plesiosaur. The fossils once again were sent to Owen, who described them as *Plesiosaurus australis*. Their age was thought to be Jurassic; after some controversy it eventually became clear that they in fact came from late Cretaceous rocks. More fossil reptiles were collected from the Waipara River exposures by Julius Haast, Provincial Geologist of Canterbury Province and first director of the Canterbury Museum. The newly established Colonial Museum in Wellington, established in 1865 by James Hector, soon joined in the collecting. As to Hood, who had returned to New Zealand in 1869 after a two years' absence, he continued to collect for Owen, and in 1869 he shipped an important collection, including a mosasaur skull, to England on board the *Matoaka*. Owen never received the fossils, however, because the *Matoaka* was lost at sea and never heard of again.

Meanwhile, both the Colonial Museum and the Canterbury Museum went on collecting on the Waipara River, and at another site 60 miles (97km) further to the north-east, Haumuri Bluff. Hector described the specimens obtained by both museums in 1874, most of them belonging to plesiosaurs and mosasaurs. Systematic collecting continued into the 1890s, after which it virtually ceased for more than 60 years. In 1890 Hector sent a shipment of thirty-nine cases of Cretaceous reptile bones to Cope. The specimens were delivered to him in Philadelphia, but they subsequently vanished and their whereabouts remain unknown (Welles and Gregg, 1971).

The early history of vertebrate palaeontology in South Africa also involved enthusiastic local collectors and identification by Richard Owen in London. Fossil bones were first found in the Permo-Triassic rocks of the Karroo by Andrew Geddes Bain (1797–1864), a Scottish engineer who was engaged in road-making in various parts of what was then known as Cape Colony. Bain himself has left a highly enjoyable account of what he termed 'the pursuit of knowledge under difficulties' (Bain, 1896). He had first become acquainted with geology in 1837 when he borrowed Lyell's *Principles of Geology* from his friend Captain Campbell. He read the book 'with avidity' and subsequently procured a copy of Buckland's *Bridgewater Treatise*, which he found an 'inestimable work'. From then on he was never without a hammer and a collecting bag, 'which conduct some charitable friends were kind enough to attribute to lunacy, when in truth it was nothing but a severe attack of lithomania' (Bain, 1896, pp. 59–60). Bain soon became acquainted with a few other Europeans who shared his interest in geology. One of them was a Mr Borcherds, Civil Commissioner at Fort Beaufort, and they went on fossil-hunting trips together, although at first the results were disappointing. Eventually, however, they found a small fossil bone on a hill a mile (1.6km) to the north of Fort Beaufort, and enthusiastically started to dig for more. Their perseverence was rewarded by the discovery of what Bain called a 'charnel-house' containing the remains of different kinds of fossil vertebrates, which they were unable to identify. Soon thereafter, Bain found at another locality a specimen which, when prepared with hammer and chisel, turned out to be a skull.

At first, he thought it could be that of a tiger, because of its 'two beautiful canine tusks', but subsequent preparation showed that there were no other teeth. Bain called the mysterious creature a 'bidental'. He extended his researches to other areas of the Karroo, and found them to 'abound in bidental and other reptiles'. His discoveries were a source of astonishment for the local population, especially the deeply religious Boers. Once, when he was excavating the skeleton of a 'huge monster' the size of a cow, a young Boer asked him what he was doing. This is how Bain (1896, pp. 62–3) reported the rather farcical conversation which ensued:

> I said in reply to the Boer's query:
> 'Don't you see that is the petrified head of a wildebeest', pointing at the same time to the open mouth containing the teeth.
> 'Alamagtig', said he, his eyes glistening in astonishment, 'how came the wildebeest in the stone?'
> 'Do you read your Bible', said I.
> 'Oh yes'.
> 'Well, did you never read that when Noah was in the ark (which contained a pair of all the different animals on the face of the earth) that one of the wildebeests jumped overboard, and before Noah could get out his life-buoy it was drowned?'
> The poor Boer at first looked rather bewildered, but not wishing to be thought deficient in Bible lore, scratching his head and looking as sheepishly as possible, said:
> 'Ja tog' (Yes, I remember).
> 'Well, then', continued I, 'you know of course that the waters covered the tops of the highest mountains, and that at that time Noah was floating above the lofty Winterberg, and the wildebeest, falling into the Fishback, became petrified there, where he was lain ever since till I took him out the day before yesterday'.
> 'Alamagtig', said the Boer again, 'het is tog wonderlijk', (it is wonderful) and saddled up his horse and rode away.

After collecting for some time Bain had accumulated the skulls of nearly forty reptiles, the greater part of which belonged to his 'bidentals', varying in size from that of a mouse to that of a rhinoceros. He attempted to present his valuable collection to a learned society in Grahamstown to become the nucleus of a museum, but the offer was turned down, the director not being interested in 'old stones'. Bain then turned toward London, and sent a long letter to Henry De la Beche, then 'Foreign Secretary' of the Geological Society of London. In it, he described his fossils as best as he could, and also gave a sketch of the geology of South Africa. He had all his specimens sent to Grahamstown for packing, and encouraged by local residents and visitors, shipped them to London. A few months later, in 1845, a letter arrived from the President of the Geological Society of London, Henry Warburton, announcing that Bain had been awarded the Wollaston Fund (amounting to about twenty guineas or £21) as a reward for his important geological and palaeontological discoveries in South Africa. The fossils had been turned over to Owen, who had had them prepared in the British Museum, and had described their anatomy in a paper read before the Society; Owen's paper had been preceded by one on the geology of South Africa prepared from Bain's letter. Owen had changed Bain's 'bidental' into the more scientific-sounding *Dicynodon*, one of the species being named after Bain. One of the most striking features of these new forms was that they exhibited mammalian characteristics.

Bain felt justifiably encouraged by the reception given to his letter and fossils in London, and decided to collect data for a geological map of South Africa. The following years were occupied, when professional duties permitted it, by extensive geological excursions on horseback or in a bullock wagon. In 1856 Bain was awarded a gift of £200 from the Queen's Privy Purse for the fossils he had collected and sent to the Geological Society. These were eventually transferred to the British Museum, and on that occasion Bain was sent £150 to cover the heavy expenses he had incurred.

Owen's description of the fossils collected by Bain (Owen, 1845) included extensive comparisons with various groups of reptiles, and although he concluded that the skull was essentially built on the lacertilian plan, he recognised that the dentition was reminiscent of mammals. Further discoveries were to confirm that what Bain had found were the remains of 'mammal-like' reptiles, but at the beginning there was much confusion as to their real affinities. In any case, the South African dicynodonts soon became famous fossils. When Prince Alfred (Queen Victoria's second son) visited South Africa in 1860, he obtained some *Dicynodon* remains (Hutchinson, 1894) which on his return he presented to Owen (the latter had often been invited by the Prince Consort to lecture before the Royal Family). Several of the leading British experts on palaeoherpetology, among them Huxley and Seeley, published on South African reptiles. In 1889 Seeley even visited Cape Colony to study fossils which had been acquired by the museums in Cape Town and Grahamstown, but he found that the best specimens had already been sent to the British Museum. He therefore decided to find more material himself, and with the help of Thomas Bain (Andrew Bain's son) he was able to collect a remarkably complete skeleton of the cotylosaur *Pareiasaurus*. The specimen was sent to London, where it was prepared and mounted in the British Museum. Thus, until the 1890s most of the important specimens of Permian and Triassic reptiles collected in South Africa were sent to London for study. This changed with the arrival of Robert Broom in South Africa. A native of Scotland, Broom (1866–1951) had studied medicine in Glasgow. He had then emigrated to Australia, where he practised medicine in the outback, and worked on living and fossil marsupials. During a visit to London in 1896, he saw South African fossil reptiles in the British Museum — and decided to settle in South Africa instead of returning to Australia. In South Africa he became a country doctor again, and started to work on mammal-like and other Permo-Triassic reptiles instead of marsupials. He collected many new specimens, and although he later became a leading authority on the australopithecines, he went on publishing on fossil reptiles until his death. With Broom began a new epoch of South African vertebrate palaeontology, when specimens no longer had to be sent to London for study, but were described in South Africa by local scientists.

In other British colonies, where the European population was much smaller, local scientific institutions developed more slowly, and until a late date fossil collecting was much more sporadic than in Canada, Australia, New Zealand or South Africa. The search for fossil vertebrates in the more remote parts of such colonies could be a perilous endeavour, as shown by the tragic fate of D.B. Pigott, a young government official in Kenya at the turn of the century. Fossil vertebrates of Miocene age had been found by G.R. Chesnaye in the Kavirondo rift valley and at a place called Karungu on the eastern shore of Lake Victoria. The specimens were shown to C.W. Hobley, the Provincial Commissioner, who sent Pigott to Karungu to search for more. As reported by Le Gros Clark and Leakey (1951, p. 1), 'unfortunately this young man fell victim to a crocodile, but not before he had collected a small series of fossils'. These were sent to the British Museum and described by C.W. Andrews. It was not until the 1930s that large-scale palaeontological expeditions explored the rich Miocene fossiliferous deposits of Kenya.

Few of the French colonies ever had a large European population, and fossil vertebrates which happened to be found there by explorers or geologists were usually sent to Paris for study. Algeria was an important exception, and there vertebrate palaeontology developed locally at an early date, mainly because of the presence of an expert on extinct vertebrates, Auguste Pomel (1821–98). Pomel had worked on the Tertiary mammals of central France with Bravard in the 1840s. At the time of the 1848 revolution he became politically active, freely expressing his radical views, and when Louis-Napoléon seized power in 1851, Pomel was denounced as a dangerous revolutionary, arrested and deported to Algeria with his family (Villot, 1957). There he first tried to earn a living as a farmer, but failed, and decided to resume his geological activities by working in a mine. In 1859 there was a general amnesty, and Pomel was at last free of periodic police control. He then engaged in local politics and at the same time started on a new scientific career. He eventually became head of the Scientific School in Algiers in 1880 and director of the Geological Survey of Algeria in 1882. He worked on many aspects of geology, but was especially interested in vertebrate palaeontology and in the 1890s published a series of long monographs on the Pleistocene mammals of Algeria. He discovered and explored a number of fossil vertebrate localities, including the famous Pleistocene site at Ternifine which later yielded *Homo erectus* remains.

While Pomel mainly worked on Pleistocene mammals, other explorers discovered older vertebrates in some parts of Algeria. In 1880, for example, fossil footprints were described by Le Mesle and Péron in Cretaceous strata in southern Algeria. Le Mesle had heard about the footprints from a French officer stationed in the area. There were local legends about them, as the Arabs thought they had been made by a giant bird. Le Mesle walked 44 km (27 miles) in one day to get to the locality, and found more than thirty footprints on the surface of a limestone bed. The rock was too hard for him to chisel out any of the prints, and he finally decided to bake some gypsum blocks which he found in the vicinity, to obtain a coarse plaster with which he made casts of some of the footprints. His interpretation was not much different from that of the local Arabs, as he thought that the three-toed footprints had been left by some colossal bird — an opinion shared by Péron. These so-called ornithichnites were in fact dinosaur footprints, the first evidence of dinosaurs to be found in North Africa.

Still further south, fossil vertebrates were found in the Sahara by military explorers. During his expedition from Algiers to the Congo via Lake Chad at the beginning of the twentieth century, for instance, Fernand Foureau collected fish and reptile remains in the Djoua region of eastern Sahara. Among them was a dinosaur vertebra described by Haug (1905), the first dinosaur bone to be reported from the Sahara. Similarly, in 1907, Captain Arnaud and Lieutenant Cortier crossed the Sahara from Algiers to the Niger and discovered crocodilian remains in the Tilemsi valley in what was then called the French Sudan (today the Republic of Mali). Later expeditions found rich early Tertiary vertebrate localities in this area.

The search for economically valuable mineral deposits in the French African colonies also resulted in the discovery of fossil vertebrates. In the 1890s, for instance, early Tertiary fish and reptile remains were found in the phosphate deposits of Tunisia (by Philippe Thomas) and Algeria (by Auguste Pomel). The phosphates were soon exploited commercially, and yielded rich vertebrate faunas.

French colonial expansion in South East Asia also led to a few interesting discoveries in the field of vertebrate palaeontology. From 1894 to 1896 a French expedition under Auguste Pavie explored Laos. Although it was ostensibly a scientific mission, one of its main aims was to bring Laos under French influence. Nevertheless, when the geologist H. Counillon studied the geology of the area around Luang Prabang on the Mekong, he was lucky enough to find a partial skull of a presumably early Triassic dicynodont (Counillon, 1896), which remains the only mammal-like reptile to have been reported from South East Asia (although the specimen, which was sent to the University of Marseilles, has since disappeared).

In some cases, the rivalry between the European colonial powers found a palaeontological expression when explorers and settlers sent the fossils they found in the disputed territories to their respective home countries. This happened in Madagascar, where the French and the British engaged in fierce competition in the later part of the nineteenth century. The first evidence of extinct vertebrates in Madagascar was in the unusual form of eggs (Lavauden, 1931). About 1830, the French captain Victor Sganzin sent notes and drawings of very large eggs to the ornithologist Verreaux. In 1834 another French traveller, Goudot, brought egg-shell fragments from Madagascar to Paris, where Paul Gervais tried to reconstruct the dimensions of the complete eggs; later discoveries showed that he had much under-estimated them. This became clear in 1850 when a Mr Abadie, the captain of a merchant ship, brought three of these eggs from the south-western coast of Madagascar to Isidore Geoffroy Saint-Hilaire. The latter called the enigmatic giant bird *Aepyornis maximus*; he thought that it was a very large ratite, but this opinion was not shared by all naturalists: Valenciennes supposed that it was a giant penguin, whereas the Italian Bianconi believed that it was a condor-like bird of prey. At the time, it was not quite clear whether *Aepyornis* was an extinct form, or was still living in remote parts of the island. In 1658 the French explorer Etienne de Flacourt had reported that an ostrich-like bird lived in some parts of Madagascar, and in the nineteenth century there were native legends about this large bird. As the exploration of the island by European travellers progressed, however, it became clear that *Aepyornis* had become extinct at a recent date. That it was not the only unexpected 'subfossil' creature to be found in Madagascar was shown by the discovery of several rich localities containing the bones of various recently extinct animals. In 1866 the French traveller Alfred Grandidier (1836–1921) found such a locality in a marsh at Ambolisatra, on the south-western coast of the island. It yielded bones of several species of *Aepyornis*, which were described in 1869 by Milne-Edwards and Grandidier, and confirmed Isidore Geoffroy Saint-Hilaire's attribution to a ratite. Together with the bones of the giant flightless birds were found those of a small hippopotamus, a crocodile and a large tortoise. After Alfred Grandidier's pioneer work, many explorers collected abundant remains of the subfossil fauna, which were sent to museums in France, England, Germany,

Austria, Norway and Switzerland. Palaeontological research in Madagascar was often made difficult by the hostility of the local population. In the 1890s the explorer Georges Muller was murdered by bandits after collecting a large number of bones at Antsirabe, in the central region of the island, a collection which eventually reached the Paris Museum (Milne-Edwards and Grandidier, 1894). Among the strangest subfossil vertebrates found in the marshes, peat-bogs and caves of Madagascar were the large extinct lemurians. Besides the collections made by French travellers, important specimens, including the type of *Megaladapis madagascariensis*, from Ambolisatra, were obtained for the British Museum by J.L. Last, who was collecting natural history specimens for Walter Rothschild (Forsyth Major, 1894). The Vienna Natural History Museum also received beautiful specimens from F. Sikora, a collector working for the Imperial Academy of Sciences, who found them in a cave at Andrahomana, near Fort Dauphin in the south-western part of the island (Lorenz von Liburnau, 1901).

The activities of explorers, travellers and missionaries in Madagasacar in the 1890s (a period when the rivalry between France and Britain for control of the island reached its highest intensity) also led to finds of much older vertebrates. In 1891, for instance, the Reverend Richard Baron of the London Missionary Society collected Jurassic reptile remains at Andranosamonta, a village in the north-western part of the island. They were sent to the British Museum for preparation and study, and turned out to be a partial skull with its lower jaw of the mesosuchian crocodile *Steneosaurus*, which was described as *S. baroni* by R.B. Newton (1893). A couple of years later the British Museum purchased a shipment of dinosaur bones collected about 20 miles (32km) inland from the north-western coast by J.L. Last. They were vertebrae, limb bones and girdle elements of large Jurassic sauropods, which Lydekker described as *Bothriospondylus madagascariensis* in 1895. French collectors did not stay idle in the search for Malagasy dinosaurs. In 1896 Depéret described sauropod bones and a theropod tooth from the Cretaceous of the Majunga region in the north-western part of the island; they had been found by Adjutant Landillon, of the French marines, who apparently collected fossils when he was not directly taking part in the French occupation of Madagascar (French troops had landed in Majunga in 1895, and within a few months Queen Ranavalona had had to accept a French protectorate. The following year Madagascar became a French colony). Also in 1896, Marcellin Boule in Paris described dinosaur bones from the Jurassic and Cretaceous of Madagascar, which had been sent by a Mr Bastard, who collected for the Paris Museum. After Madagascar became a French colony in 1896, the search for fossil vertebrates of course became mainly the work of French settlers or geologists.

Germany started to acquire colonies later than Britain and France, but the scientific exploration of the German colonies (most of them in Africa) proceeded at a rapid pace in the 1890s and until the outbreak of the First World War. Fossil fishes and reptiles found in Togo and in South West Africa were described by Ernst Stromer von Reichenbach (1871–1952), a Bavarian palaeontologist who was especially interested in the evolution and biogeographical significance of African vertebrates. However, the most spectacular result of German palaeontological activities in Africa was undoubtedly the excavation of the dinosaur localities of Tendaguru, in what was then German East Africa (later Tanganyika, today Tanzania). Large bones were first found on Tendaguru Hill by W.B. Sattler, an engineer. He understood their scientific importance, and in 1907 Eberhard Fraas, head of the Stuttgart Museum, decided to visit Tendaguru during a journey in German East Africa. This proved a difficult task: because of dysentery, Fraas was too weak to walk, and had to be carried to the locality. He nevertheless managed to collect several bones of a gigantic dinosaur. The director of the palaeontological museum of the University of Berlin, W. Branca, then resolved to excavate the late Jurassic fossil-bearing deposits on a large scale. As little money was available from the Prussian state or from the Colonial Administration, a public subscription was launched under the patronage of Duke Johann Albrecht zu Mecklenburg. The results were highly satisfying: a sum of 181,607.45 Marks was raised, which paid for the considerable expenses incurred during the first three years of the excavations (1909, 1910, 1911). Among the sponsors were German noblemen, commercial and industrial companies, banks, university professors and insurance companies. In 1912 the Prussian state finally provided funds for a fourth season of field work. Various German companies provided free miscellaneous goods, from photographic equipment to tinned soup (Branca, 1914).

The excavations started in 1909 under the direction of Werner Janensch, curator at the Berlin Museum, and his assistant Edwin Hennig (Hennig, 1912; Janensch, 1914). From the coastal city of Lindi, Janensch, Hennig, Sattler and their 160 porters reached Tendaguru after a five-day march. They found that bones and bone fragments were abundant where they had weathered out of a sandy marl, and could be picked up amid the tall grass. Trenches were dug to obtain better, unweathered material. While some of the native workers were engaged in palaeontological excavations, others were busy building a permanent camp, complete with storage buildings for bones and provisions, a large round hut for the German palaeontologists, and a whole village for the workmen (many of whom had been joined by their families). The number of workers rose from about 170 in the first year to almost 500 in the third. Providing such a large number of people with shelter, food, drinking water and medical treatment obviously caused some problems, which were solved with typical German efficiency. Ordinary workmen earned 9 rupees a month, while the more skilled supervisors and preparation assistants received between 10 and 11 rupees. Some of the native workers quickly learned to supervise the diggings, which were conducted simultaneously at several sites around Tendaguru Hill, and to carry out some preliminary preparation of the bones on the spot before they were encased in plaster jackets (or, later, in a mixture of gum-arabic and clay, which was much cheaper than plaster imported from Germany). Photographs taken by Janensch and Hennig give an impressive idea of the huge scale of the excavations, which took place only during the dry season. After careful packing in bamboo cases, the enormous bones were carried by porters to Lindi and then shipped to Germany. During the first three years, about 4,300 loads were thus carried to Lindi, and 185 tonnes of material were shipped to Berlin. In 1912 alone, 40 more tonnes were collected under the supervision of Hans Reck, corresponding to about 20,000 bones (Branca, 1914). There is little doubt that the Tendaguru digs were among the largest palaeontological excavations ever conducted, both in terms of manpower and of quantity of material collected.

The scientific results were impressive: a whole late Jurassic dinosaur fauna was collected, which at the time had no equivalent anywhere else, except in North America. In 1914 Janensch was able to publish a preliminary faunal list which was considerable enough, and grew longer as the fossils from

Tendaguru were slowly prepared in Berlin. The preparation of the often enormous and usually broken bones was an arduous and extremely time-consuming process which lasted for many years, and was delayed by the First World War and the subsequent economic difficulties of Germany. Publication of the scientific results of the Tendaguru expeditions started in 1914 and went on until the 1950s, with a long series of monographs, most of them by Janensch or by Hennig, on the various components of the fauna. Several complete skeletons of Tendaguru dinosaurs were mounted in the museum of Humboldt University in Berlin (where they fortunately survived the destruction of the Second World War). The most spectacular dinosaur found at Tendaguru was the huge *Brachiosaurus brancai*, whose skeleton dwarfs Andrew Carnegie's *Diplodocus* in the Berlin Museum. Smaller sauropods were also present, as shown by skeletons of *Dicraeosaurus*. Stegosaurs were well represented by several skeletons described by Hennig. Small ornithopods and theropods were also discovered, as well as teeth of large carnosaurs. Although dinosaurs formed the main component of the fauna (perhaps because they were preferentially collected), other vertebrates also occurred, including fishes, pterosaurs and even a mammal (represented by a toothless jaw). Just before the outbreak of the First World War, the German excavations at Tendaguru undoubtedly marked a level of colonial vertebrate palaeontology, which was never again reached. The British expeditions to Tendaguru in the 1920s did find large numbers of bones, but their scientific results were meagre by comparison with those obtained by the German palaeontologists.

In European colonies in Africa and Asia, the search for fossil vertebrates was almost solely the work of palaeontologists from the colonial power, to the exclusion of scientists from other nations. In other countries which were theoretically independent, although under the control of one or other of the colonial powers, the situation was slightly different. Egypt provides a fine example of this kind of situation: the country could not be said to be really independent as its whole administration was under British control, but it was open to explorers from all European countries. Fossil vertebrates were found in the Fayum region, west of the Nile, by the German explorer Schweinfurth, who crossed the area in 1879. The remains he found were identified as belonging to the primitive whale *Zeuglodon* by the German palaeontologist Dames. A geological survey of the Fayum province was started in 1898 under H.J.L. Beadnell. At first, only fish and crocodile remains were found. In 1901, however, the British palaeontologist C.W. Andrews went to the Fayum with Beadnell to collect recent mammals for the British Museum. The reason for Andrews's presence in Egypt at the time was that he was suffering from an illness which made the English winter weather difficult to bear, and like many in those days tried to escape it by spending the winter months in a southern country. Besides recent mammals, Andrews found fossil bones in the early Tertiary deposits of the Fayum: vertebrae of the snake *Pterosphenus*, and later abundant mammal remains. Within three weeks it was possible for Beadnell to obtain in this arid region where virtually no collection had ever been made, a good collection of largely new forms, including the primitive sirenian *Eosiren* and the early proboscideans *Barytherium* and *Moeritherium* from the late Eocene deposits, and *Palaeomastodon* from the Oligocene. As Beadnell (1905) put it, all the material was 'taken home' to be studied at the British Museum. During the following winters, until 1904, the survey was resumed, and more vertebrates were found, especially in the upper, Oligocene, levels, in which the extraordinary horned ungulate *Arsinoitherium* was unearthed. The important collections of Fayum vertebrates thus obtained by the Geological Museum in Cairo and the British Museum were described by Andrews in a long monograph published in 1906.

The rather sensational finds of Beadnell and Andrews directed the attention of many palaeontologists to the fossil-bearing formations of Egypt. Franz Nopcsa visited the Fayum in the 1900s, and collected vertebrate fossils which he presented to the British Museum. In 1907 Henry Fairfield Osborn and Walter Granger also collected specimens there for the American Museum of Natural History.

German palaeontologists seem to have been especially interested in Egypt. Eberhard Fraas, for instance, collected early whales and crocodilians from the Eocene of Mokattam, near Cairo, in the 1890s. The most active of them, however, was Ernst Stromer von Reichenbach (Dehm, 1956, 1971), whose interest in the fossil vertebrates of Egypt had been kindled by Schweinfurth's and Andrews's discoveries. During the winter of 1901–2, he made a first visit to Egypt with the geologist Blanckenhorn, and started to collect vertebrates in the Fayum area. With financial support from the Bavarian Academy of Sciences, similar winter trips to the deserts of Egypt were repeated in 1903–4 and 1910–11. Stromer first collected Tertiary vertebrates in the Neogene (Wadi Faregh, Wadi Natrun) and the Palaeogene (Fayum). He then came to the conclusion that similar but older fluviomarine facies should be found further south, and might yield remains of Cretaceous terrestrial vertebrates. He was proved right when he discovered quite abundant remains of Cenomanian reptiles in the Baharija oasis, in the western desert south of the Fayum. The fauna included sauropods, theropods (notably the bizarre *Spinosaurus* with its elongated neural spines), turtles, unusual crocodilians and early snakes, as well as abundant fishes. Stromer's important collections were sent to Munich and described by many authors (including, of course, Stromer himself) in a series of monographs on the 'results of Professor Stromer's research trips in the deserts of Egypt', published by the Bavarian Academy of Sciences between 1914 and 1936.

In Egypt, Stromer found an efficient collaborator in the person of Richard Markgraf (Stromer, 1916). Markgraf was a German from northern Bohemia who had first worked as a mason before becoming a musician. This somehow had led him to Egypt, where he fell into poverty because of ill-health. In 1897 Eberhard Fraas, who was searching for fossils in the Mokattam quarries near Cairo, taught him how to collect specimens. Markgraf then became a professional fossil collector, working mainly for such German institutions as the museums in Stuttgart, Frankfurt and Munich. Stromer took him to Wadi Natrun, the Fayum localities and Baharija. Despite an increasingly severe (and unspecified) illness, Markgraf continued to collect for German museums, and also for the American Museum, until the First World War severed his relations with Germany and ruined his business. He died in the Fayum, completely destitute, in 1916. One of his most spectacular discoveries had been that of primate jaws from the Oligocene which he sent to Stuttgart, and which were described as *Propliopithecus*, *Parapithecus* and *Moeripithecus* by Max Schlosser.

Very few countries in Africa and Asia managed to remain more or less independent during the great phase of European colonial expansion of the late nineteenth century. In such countries palaeontological exploration was usually done by a few

European travellers often under difficult conditions. China provides a good example of such a situation: although several European powers acquired concessions in the main ports, or even small colonies along the coast, the interior of the country was not much penetrated by European influence. Because of the use of fossil bones and teeth in traditional Chinese medicine (see Chapter 1), however, specimens from fossil localities in the interior were available in quite large quantities in the drugstores of the main cities, and Europeans were soon able to obtain and study such fossils. As early as 1853, T. Davidson had described a small collection of fossils which had been purchased in a drug warehouse in Shanghai by a Mr W. Lockhart, who had sent them to a Mr Hanbury in England, who in turn had presented them to the British Museum. Most of the specimens were Devonian brachiopods (also thought by the Chinese to have curative virtues), but there were also some mammalian remains. They were examined by Waterhouse, who identified teeth of a rhinoceros, a small ruminant, two species of stags, a bear, and *Hippotherium* (a synonym of *Hipparion*, a frequent fossil in the Chinese Neogene). These teeth were stated to come from the Provinces of 'Shen-si and Shan-si' (Shaanxi and Shanxi in the modern Chinese transliteration). Somewhat later, an 'elephant' tooth was sent from Shanghai to London, followed by more teeth provided by Robert Swinhoe, the British Consul in Formosa (Taiwan). These had been found in a cave near the city of Chungking in the Province of Sichuan. They were studied by Owen (1870), who identified the proboscidean *Stegodon* (already known from the fossils collected in Burma by Crawfurd), *Hyaena, Rhinoceros* and *Tapirus*.

Even when Europeans were able to travel in the interior provinces of the Chinese Empire, they had few opportunities for palaeontological excavations, and usually got their fossil vertebrates from Chinese traders. Thus, in 1885 the German palaeontologist Ernst Koken described a number of Pleistocene mammal remains which had been acquired in China by the famous explorer Ferdinand von Richthofen. Richthofen had seen large quantities of such fossils being brought down the Yangzi River on barges, to be sold to drugstores. The bones and teeth were usually broken, and Richthofen had selected the best he could find. They were said to come from caves in Yunnan.

In a few instances, nevertheless, European explorers may have been able to collect fossil vertebrates *in situ*. In 1872, for instance, Gaudry described a small collection of Pleistocene mammal remains (belonging to an elephant, a woolly rhinoceros, a horse, an ox and a deer) which had been found at a place called Suen Hoa Fou in inner Mongolia, north of Beijing, by Father David. Father David was a French priest who travelled extensively in China in the 1860s and 1870s, and made several important zoological discoveries (including that of the deer which now bears his name, and that of the giant panda). To judge from Gaudry's very brief indications, it seems that Father David collected the Suen Hoa Fou fossils himself.

Nevertheless, most of the fossil mammal remains which reached Europe from China at the end of the nineteenth and the beginning of the twentieth centuries came from drugstores. At the turn of the century a large collection of such fossils was sent to the Bavarian State Collection in Munich by Dr K. Haberer, who had obtained them in the drugstores of several large Chinese cities, including Shanghai, Tianjin and Beijing (Schlosser, 1902). The fossils came mainly from Neogene deposits, and an idea of the size of the collection is given by the fact that it included nearly 1,000 teeth of *Hipparion*! Until well into the twentieth century, traditional Chinese drugstores remained a major source of Pleistocene and Neogene fossil mammals — with the drawback that the specimens were usually broken and that little was known about their precise geographical and geological origin.

After the suppression of the Boxer uprising of 1900 by armies from western Europe, Russia, the United States and Japan, China had to become more open to foreign influence, and geologists from various countries at last had the opportunity to explore the country more thoroughly. This resulted in the first discoveries of Chinese dinosaurs. According to Dong (1985), 'dragon bones' from Sichuan mentioned in a book of the western Jin Dynasty (AD 265–317) were probably dinosaur bones, but this record can hardly be called a scientific document. In the southwestern part of China, the first dinosaur remains to be found were part of a femur and a carnosaur tooth, discovered by the American geologist Loudback during a geological survey of Sichuan (still one of the most productive regions of China in terms of dinosaurs) conducted between 1913 and 1915 (Dong, 1985). Loudback gave the fossils to the University of California, where C.L. Camp described them.

The Cretaceous dinosaur localities of Shandong were discovered in 1913 when a French priest, R. Mertens, found several pieces of dinosaur bones, which were taken to Beijing. The locality was rediscovered in 1922; a partial skeleton of the sauropod *Euhelopus* was excavated and sent to Upsala.

One of the earliest and most spectacular discoveries of dinosaur remains in China took place in Manchuria, which at the beginning of the twentieth century was under Russian influence (although this diminished after Russia's defeat in the Russo-Japanese war of 1905). In 1902 a few very large bone fragments found by a fisherman were interpreted as those of a mammoth by a Russian colonel by the name of Manakin. They came from a site on the Chinese bank of Heilongjiang (Black Dragon River), better known to Europeans as the Amur River, which there forms the boundary between Russia and China. Manakin's report attracted some attention, and in 1914 more remains were found by the geologist A.N. Krishtofovich. Another Russian geologist, V.P. Renngarten, was sent to investigate the area more thoroughly in 1915–16, and finally an excavation was conducted in 1916 and 1917 by N.P. Stepanov for the Russian Geological Committee. The dinosaur remains thus unearthed were sent to St Petersburg (which soon after became Leningrad), where they were prepared during the years of the revolution and the civil war. In 1924 a much restored hadrosaur skeleton was finally mounted in the museum of the Geological Committee. A.N. Riabinin described it as *Mandschurosaurus amurensis* in 1930 (Dong, 1985). Thus, the first reasonably well-preserved dinosaur skeleton ever found in China left the country soon after its discovery. In China, as in many other Asian and African countries, it was not until local vertebrate palaeontologists became active on a sufficiently large scale that important national or regional collections could be built up. This began in the late 1920s and 1930s in China, and rather later in other countries.

10

Epilogue: A Brief Review of Developments since 1914

The lives of vertebrate palaeontologists were as much affected by the First World War as those of many others. At least one well-known expert on fossil vertebrates died as a result of war-related activities: Armand Thevenin, Marcellin Boule's assistant at the Paris Museum, who had published valuable studies on fossil amphibians and reptiles, died in 1918 after experimenting on poison gas for the French army. Many had their careers interrupted by the war. In 1914, for instance, the German Fritz Berckhemer was doing post-doctoral research at Columbia University; when the war broke out he tried to return to his country on board a Dutch ship, but somehow was captured by the French and interned for the duration of the war (Dehm, 1956). He later became the head of the Stuttgart Natural History Museum and worked on various groups of fossil vertebrates from southern Germany. As to Franz Nopcsa, he spent most of the war spying in Albania and Romania for the Austro-Hungarian Empire.

In some cases, fossil vertebrates were destroyed or lost because of military activities. In 1916, for example, dinosaur skeletons collected on the Red Deer River by Charles Sternberg were shipped to London on board the *Mount Temple* — but never reached the British Museum, for the ship was torpedoed in the Atlantic by a German submarine (Colbert, 1968). However, losses were to some extent balanced by unexpected discoveries during military operations. The best example is that of the late Miocene mammals of the Salonika region in northern Greece (Bouvrain and de Bonis, 1984). After the failure of the Dardanelles expedition in 1915, Allied forces were landed in Macedonia, and among the French officers who commanded troops in this area was Camille Arambourg, a vertebrate palaeontologist in civilian life. When his soldiers began to dig trenches some 30 kilometres (19 miles) north-west of Salonika, they soon came across abundant fossil bones — and Arambourg was able to occupy his men during the dull spells of trench warfare by having them excavate the rich late Miocene deposits of the region. The abundant mammal remains thus recovered were then sent to Paris and studied by Arambourg and Piveteau after the war.

Still more unexpectedly, vertebrate palaeontology was even used for propaganda purposes. In 1916 Marcellin Boule contributed an article on palaeontology to a collection of papers published under the title *Les Allemands et la Science* (The Germans and Science). The general purpose of the book was to show that the much vaunted German science was inferior to French and Anglo-Saxon science, by asking the opinion of various eminent French scientists (and it must be said that most of them enthusiastically accepted this rather unsavoury task). Boule's paper (1916) on 'war and palaeontology' went beyond the usual deprecating remarks about 'obscure and heavy' German thought, to establish a parallel between the evolution of the German people and that of animal lineages, which was revealing of both his feelings toward the Germans and his ideas on evolution (both being rather typical of the period). The Germans, like the dinosaurs and the sabre-toothed tigers, had become over-specialised in the direction of brute strength, and the laws of evolution clearly showed that this was leading them to extinction. All of this, moreover, was simply a result of 'the natural laws of our great Lamarck on the influence of the environment and of heredity': the French had evolved in beautiful and clear landscapes, whereas the Germans had developed amidst the 'foggy marshes' of central Europe. There was therefore no doubt that the French soldiers (and their allies) would ultimately destroy the German monster, 'whose evolution had taken a wrong turn'.

Developments in the field of vertebrate palaeontology since the First World War can be divided into a few categories: discovery of new localities and, more generally, progress in the palaeontological exploration of the world, progress of palaeontological techniques and conceptual progress.

In the 1920s and 1930s economic conditions in Europe were

not favourable to large-scale palaeontological expeditions abroad, especially in the defeated countries of central Europe. However, quite ambitious projects could still be launched inside some European countries. An example directly linked to the economic difficulties of the post-war years is that of Othenio Abel's excavations in the Drachenhöhle, the famous fossiliferous cave near Mixnitz in Styria. After the dismemberment of Austria-Hungary in 1919, the German-speaking part of the former Hapsburg Empire found itself in a situation of complete economic breakdown. Among many other problems, no money was available to import the phosphates which were needed for agricultural recovery, and the Austrian government had to resort to a traditional source of fertiliser, the Pleistocene cave fillings. In 1920 the exploitation of the phosphate-rich soil of the Drachenhöhle was begun, under the scientific supervision of Abel (Abel, 1932). The excavations yielded 3,500 tonnes of phosphoric acid, which were used for agricultural purposes, and thousands of bones of the cave bear, which were sent to the Palaeontological Institute of the University of Vienna. Their study provided a wealth of information on the palaeobiology of the cave bear.

Another example of spectacular palaeontological researches under difficult economic circumstances is provided by the excavations conducted by Friedrich von Huene and his team from the University of Tübingen at Trossingen, in south-western Germany. There, from 1921 to 1923, a rich late Triassic locality was systematically excavated, and many skeletons of the prosauropod dinosaur *Plateosaurus* were unearthed (Colbert, 1968; Weishampel, 1984).

In the 1920s American institutions were in a better position to organise palaeontological excavations, not only in the United States, but also in foreign countries. The American Museum of Natural History thus sent Barnum Brown to several important localities in Europe (Samos) and Asia (Burma). Certainly, the most spectacular palaeontological expeditions of the 1920s were the Central Asiatic Expeditions of the American Museum of Natural History in China and Mongolia, under the leadership of Roy Chapman Andrews (Andrews, 1932), with Walter Granger as chief palaeontologist. The initial aim of the expeditions was to search for remains of early man (central Asia being considered at the time as a likely cradle for mankind, as well as for many other groups of mammals). Instead of human remains, however, the

American expeditions found Cretaceous mammals and dinosaurs (the latter with their eggs), and rich Tertiary faunas — until political turmoil in China brought the expeditions to an end.

After the Second World War (which incidentally caused the destruction by bombing of several valuable fossil vertebrate collections, notably in Munich, Caen and Le Havre), important palaeontological expeditions to hitherto 'unexplored' regions, or to especially interesting areas, were resumed. Most parts of the world have now been visited by vertebrate palaeontologists, and a detailed list of the major expeditions of the last forty years would require many pages. The search for fossil vertebrates in the Gobi Desert of Mongolia was continued by Russian expeditions in the 1940s and 1950s, Polish-Mongolian expeditions in the 1960s and early 1970s, and Russian-Mongolian expeditions since then. Rich dinosaur localities have been discovered in the Sahara by French palaeontologists. Even the polar regions have yielded important and sometimes abundant vertebrate remains: Devonian vertebrates in Greenland and Spitzbergen, Triassic amphibians and reptiles in Antarctica, Eocene mammals on Ellesmere Island in the Canadian Arctic. The much publicised palaeoanthropological expeditions of the 1970s to East Africa have also resulted in a much better understanding of the evolution of vertebrate faunas in the Neogene and Pleistocene of Africa. Even such countries as Thailand, which had not been much visited by European or American palaeontologists because they had managed not to be colonised, are now yielding interesting vertebrate faunas, thanks to the support of local geological surveys and international co-operation. Some of the most fascinating palaeontological discoveries of recent years have come from China, where active teams of Chinese vertebrate palaeontologists are exploring remarkable fossiliferous formations ranging in age from the early Palaeozoic to the Pleistocene. It should also be mentioned that in many developing countries fossil vertebrates are now considered as part of the national heritage, and protected as such.

Vertebrate palaeontology has benefited both from general technical progress and from the development of specialised procedures which have permitted a better investigation of fossils. Technical innovations of broader scope which have found useful applications in vertebrate palaeontology include X-ray photography, the development of scanning electron microscopy and the increasing use of computers, to mention but a few examples.

The specialised palaeontological techniques which have made much progress in the twentieth century are mainly aimed at a more complete recovery of fossil documents and at a better preparation of specimens. Important progress has taken place, for instance, in the collecting of small vertebrate remains: the washing and sieving methods pioneered by Lartet in the 1830s have been much amplified and improved since the Second World War, with the use of various separation techniques (dense liquids, interfacial method, etc.). The washing and screening of large amounts of fossiliferous sediment has now become common practice, and has resulted in spectacular progress in the knowledge of many groups of small mammals which were previously known only from a few specimens. Mesozoic mammals, for instance, are now known from thousands of specimens from many localities instead of a few isolated finds. Our present knowledge of rodent evolution, with its important biostratigraphic implications, is also a result of improved collecting techniques. Improvements in the field of preparation techniques include the use of chemicals (mainly acids) to get rid of matrix; such procedures in many cases are much less damaging to the fossils than was the old mechanical preparation, and have allowed the accurate cleaning of very delicate specimens. Another technique which has contributed greatly to a better knowledge of the internal anatomy of some fossil vertebrates (notably Palaeozoic agnathans and fishes) is serial sectioning, which was pioneered in the first decades of the century by the British palaeontologist W.J. Sollas. Although this technique results in the destruction of the original specimen, it yields information which is difficult to obtain otherwise. Improved mounting techniques have also resulted in a more life-like and spectacular presentation of skeletons of fossil vertebrates in modern museums.

Finally, conceptual advances have been numerous and important, and can be mentioned only briefly here. In the field of evolutionary theory, the contribution of vertebrate palaeontology (and of palaeontology as a whole) to the development of the modern neo-Darwinian synthesis has been relatively limited, for the simple reason that palaeontology, as a historical science, has little to say about the mechanism of change — even though the reconstruction of phylogenies is still one of the major aims of palaeontological research. The major elements of the synthetic theory have been provided by other disciplines such as genetics, molecular biology and ecology. Most vertebrate palaeontologists seem to have realised that attempts at deducing general, quasi-mathematical laws from the fossil record, as was enthusiastically done at the turn of the century, are doomed to failure. While it is now recognised that the mechanisms of organic evolution lie largely (if not completely) beyond the reach of palaeontology, it is generally accepted that the fossil record provides useful data concerning the pattern of evolutionary change, as illustrated by George Gaylord Simpson's classic *Tempo and Mode in Evolution* (1944). Current controversies about 'punctuated equilibria' versus 'phyletic gradualism' are very largely based on palaeontological evidence — and mainly demonstrate that the evidence is far from unequivocal, despite attempts at quantification. An interesting recent development of studies on the 'tempo and mode of evolution' has been the resurrection of catastrophist hypotheses to explain some of the mass extinctions in the history of vertebrates (and of life in general). Although there is nothing supernatural about the new catastrophism, the idea of relatively brutal extinction on a large scale (possibly caused by extra-terrestrial events) has proved repugnant to some palaeontologists, whereas others have accepted it with enthusiasm — which has led to a considerable amount of controversy. The historical data provided by vertebrate palaeontology have also been useful in the development of new methods of phylogenetic reconstruction: dates of divergence of groups deduced from the fossil record have provided a more or less accurate chronological framework for the calibration of the 'molecular clock'.

Many twentieth-century vertebrate palaeontologists have felt that a more fruitful approach was the detailed description of morphological structures in fossil forms which were thought to be of especial evolutionary significance. Such extremely detailed studies were made possible by some of the improved preparation techniques mentioned above, and the so-called 'Swedish school' of vertebrate palaeontology (followed by specialists in other countries) has carried such investigations to a considerable degree of sophistication.

Other palaeontologists, however, have attempted to interpret fossil vertebrates in relation to their environment, and to use them for purposes of biochronological correlation and palaeogeographical reconstruction. Microvertebrates, because of their considerable abundance in some deposits, have proved particularly useful for some types of palaeoecological reconstructions, as well as for biostratigraphy. Accurate vertebrate-based biochronological scales have been developed for the continental Cenozoic of Eurasia, North America and South America, and attempts are being made to establish similar scales for other regions and other periods.

Within the last few years spectacular progress has been achieved in the reconstruction of the palaeobiogeographical history of many groups of vertebrates. As indicated elsewhere in this book, the problems posed by the past distribution of vertebrates have been a challenge to palaeontologists since the eighteenth century. The most significant progress in the history of palaeobiogeography has of course been the acceptance of continental drift. Alfred Wegener (1880–1930) had partly been led to the concept of continental drift by papers on palaeobiogeography (Schwarzbach, 1980), and he made much use of palaeontological evidence in his great work *Die Entstehung der Kontinente und Ozeane* (The origin of continents and oceans — fourth German edition 1929). The response of vertebrate palaeontologists to Wegener's hypothesis of moving continents was variable: although some (including Franz Nopcsa) were enthusiastic, many were reluctant or sceptical (Buffetaut, 1980). The refusal of most geologists and geophysicists to accept continental drift resulted in its rejection by the vast majority of vertebrate palaeontologists. While in the 1950s some still accepted the idea of 'continental bridges' across vast oceanic spaces, others, such as George Gaylord Simpson, tried to explain the palaeobiogeographical history of vertebrates within a geographical framework not much different from the present one. Both approaches were doomed to failure, and when the emergence of plate tectonics in the 1960s (Muir Wood, 1985) led to the final acceptance by the geological community of continental displacement, it soon became clear that many questions raised by the past distribution of vertebrates could be answered more easily in a mobilist framework. This is not to say that all problems have been solved; on the contrary, the reconstruction of the palaeobiogeographical history of vertebrates is fraught with many challenging difficulties, which make it one of the most active (and attractive) branches of vertebrate palaeontology.

No definitive conclusion can be given to what is, fortunately, an unfinished story. The number and importance of the various gaps in our knowledge of vertebrate evolution in time and space clearly demonstrate that much remains to be done in the field of vertebrate palaeontology, despite the enormous amount of work which has been done since the beginning of the nineteenth century. Future developments cannot really be foreseen, all the more because palaeontological discovery is largely a matter of chance, as should be apparent from the preceding account. But whatever in the years to come the evolution of vertebrate palaeontology will be, a consequence of what has already been achieved is worth stressing: two hundred years ago almost nothing was known of fossil vertebrates; the possible former existence of now extinct species was accepted by very few scientists, and the concept of a long succession of vanished faunas gradually leading to the present one would still take many years to emerge. Today, the mammoth has become the trademark of a French chain of supermarkets, and most children in developed countries are more familiar with dinosaurs than with many living animals. Such examples of course do not indicate that the major discoveries of vertebrate palaeontology are known and understood by all (after all, in those of our societies which pride themselves on their scientific achievements, a reasonable amount of scientific knowledge is not generally considered as a necessary part of the cultural background of an educated person). They do show, however, that some of the long-extinct creatures of the past revealed by vertebrate palaeontology have now become familiar to large numbers of people well outside scientific circles. This is not a negligible achievement for what is sometimes considered as a rather abstruse branch of scientific knowledge.

Bibliography

Abel, O. (1912) *Grundzüge der Palaeobiologie der Wirbeltiere*, Stuttgart
—— (1914) 'Paläontologie und Paläozoologie' in R. Hertwig and R. von Wettstein (eds), *Die Kultur der Gegenwart*, Berlin and Leipzig, 3, 4, 303–95
—— (1917) *Allgemeine Paläontologie*, Berlin and Leipzig
—— (1925) *Geschichte und Methode der Rekonstruktion vorzeitlicher Wirbeltiere*, Jena
—— (1927) *Lebensbilder aus der Tierwelt der Vorzeit*, 2nd edn, Jena
—— (1932) 'La vie des animaux de l'époque glaciaire dans la Caverne des Dragons, à Mixnitz, en Styrie', *La Terre et la Vie*, 2, 2, 1–24
—— (1939a) *Vorzeitliche Tierreste im deutschen Mythus, Brauchtum und Volksglauben*, Jena
—— (1939b) *Tiere der Vorzeit in ihrem Lebensraum*, Berlin
Abildgaard, P.C. (1796) *Kort Beretning om det Kongelige Natural-cabinet i Madrid, med en Beskrivelse over et gigantisk Skelet af et nut ubekiendt Dur, som er opgravet i Peru og bevares i dette Museum*, Copenhagen
Adam, K.D. (1971) 'Ichthyosaurier aus dem schwäbischen Jura. Ein Beitrag zur Forschungsgeschichte' in *Katalog zur Ausstellung in der Girokasse*, Stuttgart, pp. 12–16
Adams, A.L. (1877) 'On a fossil saurian vertebra, *Arctosaurus osborni*, from the Arctic regions', *Proc. Roy. Irish Acad.*, 2, 177–9
Agassiz, L. (1833–44) *Recherches sur les poissons fossiles*, Neuchâtel and Soleure
—— (1844) *Monographie des poissons fossiles du Vieux Grès Rouge ou système Dévonien (Old Red Sandstone) des Iles britanniques et de Russie*, Neuchâtel
Albritton, C.C. (1980) *The Abyss of Time*, San Francisco
Alléon Dulac, J.L. (1763) *Mélanges d'histoire naturelle*, Lyon
Amalitzky, V.P. (1900) *Sur les fouilles de 1899 de débris de Vertébrés dans les dépôts permiens de la Russie du Nord*, Warsaw
Ameghino, F. (1906) 'Les formations sédimentaires du Crétacé supérieur et du Tertiaire de Patagonie', *An. Mus. Nac. Buenos Aires*, 3, 8, 1–568
Andrews, C.W. (1906) *A Descriptive Catalogue of the Tertiary Vertebrata of the Fayum, Egypt*, London
—— (1910–13) *A Descriptive Catalogue of the Marine Reptiles of the Oxford Clay*, London
Andrews, R.C. (1932) *The New Conquest of Central Asia*, New York
Andrews, S.M. (1982) *The Discovery of Fossil Fishes in Scotland up to 1845*, Edinburgh
Archer, M. and Hand, S. (1984) 'Background to the search for Australia's oldest mammals' in M. Archer and G. Clayton (eds), *Vertebrate Zoogeography and Evolution in Australasia*, Carlisle, 517–65

200

Archiac, A. d' (1864) *Introduction à l'étude de la paléontologie strati-graphique*, Paris

Baier, J.J. (1708) *Oryctographia Norica*, Nuremberg

Bain, A.G. (1896) 'Reminiscences and anecdotes connected with the history of geology in South Africa, or the pursuit of knowledge under difficulties', *Trans. Geol. Soc. South Africa*, 2, 59–75

Balzac, H. de (1831) *La peau de chagrin*, Paris

Barale, G., Bernier, P., Bourseau, J.P., Buffetaut, E., Gaillard, C., Gall, J.C. and Wenz, S. (1985) *Cerin, une lagune tropicale au temps des dinosaures*, Lyon

Bassett, M.G. (1982) *'Formed stones', Folklore and Fossils*, Cardiff

Beadnell, H.J.L. (1905) *The Topography and Geology of the Fayum Province of Egypt*, Cairo

Belcher, E. (1855) 'Notice of the discovery of *Ichthyosaurus* and other fossils in the late Arctic Searching Expedition, 1852–54', *Rep. Brit. Ass. Adv. Sc.*, 25, 79

Benton, M.J. (1982) 'Progressionism in the 1850s: Lyell, Owen, Mantell and the Elgin fossil reptile *Leptopleuron* (*Telerpeton*)', *Arch. Nat. Hist.*, 11, 1, 123–36

Biraben, M. (1968) *German Burmeister. Su vida. Su obra*, Buenos Aires

Blainville, H. de (1835) 'Mémoire sur les ossements fossiles attribués au prétendu géant Theutobochus, roi des Cimbres', *Nouv. Ann. Mus. Hist. Nat. Paris*, 4, 37–73

—— (1837) 'Sur les ossements fossiles attribués au prétendu géant Theutobochus', *C. R. Acad. Sc. Paris*, 4, 633–4

—— (1838a) 'Doutes sur le prétendu didelphe fossile de Stonefield', *C. R. Acad. Sc. Paris*, 4, 402–18

—— (1838b) 'Nouveaux doutes sur le prétendu didelphe de Stones-field', *C. R. Acad. Sc. Paris*, 7, 727–36

Blot, J. (1969) 'Les poissons fossiles du Monte Bolca', *Mem. Mus. Civ. Stor. Nat. Verona*, 2, 1–525

Blumenbach, J.F. (1799) *Handbuch der Naturgeschichte*, 6th edn, Göttingen

—— (1803) 'Specimen archaeologiae telluris terrarumque inprimis Hannoveranarum', *Comment. Soc. Gotting.*, 15, 132–56

—— (1806) *Beiträge zur Naturgeschichte*, 1. Theil, 2nd edn, Göttingen

Bondesio, P. (1977) 'Cien años de paleontologia en el Museo de La Plata' in *Obra del Centenario del Museo de La Plata*, 1, 75–87

Boule, M. (1896) 'Note préliminaire sur les débris de dinosauriens envoyés au Muséum par M. Bastard', *Bull. Mus. Hist. Nat. Paris*, 2, 347–51

—— (1916) 'La guerre et la paléontologie' in G. Petit and M. Leudet, *Les Allemands et la Science*, Paris

Bourdier, F. (1969) 'The campaign for palaeontological evolution' in C.J. Schneer (ed.), *Toward a History of Geology*, Cambridge and London

Bourseau, J.P., Buffetaut, E., Barale, G., Bernier, P., Gaillard, C., Gall, J.C. and Wenz, S. (1984) 'La carrière des calcaires litho-graphiques de Cerin (Ain, commune de Marchamp). Vie et extinction d'une exploitation communale sur un gisement paléontologique célèbre', *Nouv. Arch. Mus. Hist. Nat. Lyon*, 22 (suppl.), 21–30

Bouvrain, G. and Bonis, L. de (1984) 'La faune européenne il y a dix millions d'années', *Pour la Science*, 78, 38–45

Bowler, P.J. (1976) *Fossils and Progress: Paleontology and the Idea of Progressive Evolution in the Nineteenth Century*, New York
—— (1984) *Evolution. The History of an Idea*, Berkeley, Los Angeles, London

Brach, J.P. (1984) 'Les sources astronomiques de G. Cuvier dans le *Discours sur les révolutions du globe*' in E. Buffetaut, J.M. Mazin and E. Salmon (eds), *Actes du Symposium Paléontologique G. Cuvier*, Montbéliard, 53–8

Branca, W. (1914) 'Allgemeines über die Tendaguru-Expedition', *Arch. Biontol.*, 3, 1, 1–13

Bronn, H.G. (1861) *Essai d'une réponse à la question de prix proposée par l'Académie des Sciences pour le concours de 1853, et puis remise pour celui de 1856*, Paris

Brookes, R. (1763) *The Natural History of Waters, Earths, Stones, Fossils and Minerals With Their Virtues, Properties and Medicinal Uses: to Which is Added the Method in Which Linnaeus has Treated These Subjects*, London

Brown, B. (1926) 'Is this the earliest known fossil collected by man?', *Natural History*, 26, 535

Buckland, W. (1824) 'Notice on the *Megalosaurus*, or great fossil lizard of Stonesfield', *Trans. Geol. Soc. London*, 2, I, 390–6
—— (1828) 'Geological account of a series of animal and vegetable remains and of rocks, collected by J. Crawfurd, Esq. on a voyage up the Irawadi to Ava, in 1826 and 1827', *Trans. Geol. Soc. London*, 2, 3, 377–92
—— (1836) *Geology and Mineralogy Considered with Reference to Natural Theology*, London

Buffetaut, E. (1979) 'A propos du reste de dinosaurien le plus ancien-nement décrit: l'interprétation de J.B. Robinet (1768)', *Histoire et Nature*, 14, 79–84
—— (1980) 'Alfred Wegener et la théorie de la dérive des continents: un aperçu historique', *Bull. Soc. géol. Normandie*, 67, 4, 7–19
—— (1983) 'La paléontologie des Vertébrés mésozoïques en Normandie du 18e siècle à nos jours: un essai historique', *Actes Mus. Rouen*, 2, 39–59
—— (1985) 'The strangest interpretation of *Archaeopteryx*' in M.K. Hecht, J.H. Ostrom, G. Viohl and P. Wellnhofer (eds), *The Beginnings of Birds*, Eichstätt, pp. 369–70
—— and Wouters, G. (1978) 'Le centenaire des Iguanodons de Bernis-sart', *La Recherche*, 88, 390–2

Buffon, G.L. Leclerc, comte de (1749) *Histoire naturelle, générale et particulière, avec la description du Cabinet du Roy*, 1, Paris
—— (1778) *Histoire naturelle, générale et particulière, contenant les époques de la nature*, 9, Paris

Burnet, T. (1680) *Telluris Theoria Sacra*, London

Camper, A. (1799) 'Over den oorsprong der uitgedolven beenderen van den St. Pietersberg, bÿ Maestricht', *Natuurk. Verh. Holland. Maatsch. Wet.*, *1*, 169–98

Camper, P. (1786) 'Conjectures relative to the petrifications found in St Peter's Mountain near Maestricht', *Phil. Trans. Roy. Soc. London*, *76*, 443–56

Canéto (1873) *Le Dinothérium gigantesque dans les dépôts de notre Sud-Ouest, spécialement dans le Département du Gers*, Auch.

Caumont, A. de (1828) 'Essai sur la topographie géognostique du département du Calvados', *Mém. Soc. Linn. Normandie*, 59–366

Chambers, R. (1844) *Vestiges of the Natural History of Creation*, London

Chapman, W. (1758) 'An account of the fossile bones of an allegator, found on the sea-shore, near Whitby in Yorkshire', *Phil. Trans. Roy. Soc. London*, *50*, 688–91

Charig, A.J. (1979) *A New Look at the Dinosaurs*, London

Colbert, E.H. (1968) *Men and Dinosaurs*, New York

—— (1970) *Fossils of the Connecticut Valley. The Age of Dinosaurs Begins*, Hartford

Collini, C. (1784) 'Sur quelques zoolithes du Cabinet d'Histoire Naturelle de S.A.S.E. Palatine et de Bavière', *Acta Acad. Theodoro-Palatinae*, *5*, 58–103

Collinson, P. (1768) 'An account of some very large fossil teeth, found in North America', *Phil. Trans. Roy. Soc. London*, *57*, 464–7

Colonna, F. (1616) *De glossopetris dissertatio*, Rome

Cope, E.D. (1884) *The Vertebrata of the Tertiary Formations of the West*, Washington

Counillon, H. (1896) 'Documents pour servir à l'étude géologique des environs de Luang Prabang (Cochinchine)', *C. R. Acad. Sc. Paris*, *123*, 1330–3

Cowper Reed, F.R. (1897) *A Handbook to the Geology of Cambridge-shire*, Cambridge

Croizet J.B. and Jobert, A. (1828) *Recherches sur les ossemens fossiles du Département du Puy-de-Dôme*, Paris

Cuvier, G. (1796a) 'Notice sur le squelette d'une très-grande espèce de quadrupède inconnue jusqu'à présent, trouvé au Paraguay, et déposé au cabinet d'Histoire naturelle de Madrid', *Magasin encyclopédique*, *1*, 303–10

—— (1796b) 'Mémoire sur les espèces d'éléphans vivantes et fossiles', *Magasin encyclopédique*, *1*, 440–5

—— (1800) 'Sur une nouvelle espèce de crocodile fossile', *Bull. Soc. philomath. Paris*, *2*, 159

—— (1801) 'Extrait d'un ouvrage sur les espèces de quadrupèdes dont on a trouvé les ossemens dans l'intérieur de la terre', *J. Phys. Chim. Hist. Nat.*, *52*, 253–67

—— (1804a) 'Sur les espèces d'animaux dont proviennent les os fossiles répandus dans la pierre à plâtre des environs de Paris. 1er Mémoire: Restitution de la tête', *Ann. Mus. Hist. Nat. Paris*, *3*, 275–303

—— (1804b) 'Mémoire sur le squelette presque entier d'un petit quadrupède du genre des sariques, trouvé dans la pierre à plâtre des environs de Paris', *Ann. Mus. Hist. Nat. Paris*, *5*, 277–92

—— (1812) *Recherches sur les ossemens fossiles de quadrupèdes, où l'on rétablit les caractères de plusieurs espèces d'animaux que les révolutions du globe paroissent avoir détruites*, Paris

—— (1834–6) *Recherches sur les ossemens fossiles, où l'on rétablit les caractères de plusieurs animaux dont les révolutions du globe ont détruit les espèces*, Paris

—— and Brongniart, A. (1810) 'Essai sur la géographie minéralogique des environs de Paris', *Mém. Cl. Sc. math. phys. Inst. Impér. France*, *1*, 1–274

Darwin, C.R. (1845) *Journal of Researches into the Natural History and Geology of the Various Countries Visited by H.M.S. Beagle*, London

—— (1859) *On the Origin of Species by Means of Natural Selection: or the Preservation of Favoured Races in the Struggle for Life*, London

Daubenton, L. (1764) 'Mémoire sur des os et des dents remarquables par leur grandeur', *Mém. Acad. Roy. Sc. Paris* (1762), 206–29

Daubrée, A. (1858) 'Découverte de traces de pattes de quadrupèdes dans le grès bigarré de Saint-Valbert près Luxeuil (Haute-Saône)', *Mém. Soc. Hist. Nat. Strasbourg*, *5*, 1–8

Davidson, T. (1853) 'On some fossil brachiopods, of the Devonian age, from China', *Q. Jl Geol. Soc. London*, *9*, 353–9

Decrouez, D. (1981) 'Bernissart et ses Iguanodons', *Musées de Genève*, *219*, 2–8

Defay (1783) *La nature considérée dans plusieurs de ses opérations, ou mémoires et observations sur diverses parties de l'histoire naturelle, avec la minéralogie de l'Orléanois*, Paris

Dehm, R. (1956) 'Rückblick', *Paläont. Z.*, *30*, 3/4, 215–25

—— (1971) 'Professor Dr Ernst Freiherr Stromer von Reichenbach, Lebensdaten und Schriftenverzeichnis', *Mitt. Bayer. Staatssamml. Paläont. hist. Geol.*, *11*, 3–10

De la Beche, H.T. and Conybeare, W.D. (1821) 'Notice of the discovery of a new fossil animal, forming a link between the *Ichthyosaurus* and crocodile, together with general remarks on the osteology of the *Ichthyosaurus*', *Trans. Geol. Soc. London*, *5*, 558–94

Delair, J.B. and Sarjeant, W.A.S. (1975) 'The earliest discoveries of dinosaurs', *Isis*, *66*, 5–25

Depéret, C. (1896) 'Note sur les dinosauriens sauropodes et théropodes du Crétacé supérieur de Madagascar', *Bull. Soc. Géol. France*, *24*, 176–94

—— (1904) 'Sur les caractères et les affinités du genre *Chasmotherium* Rütimeyer', *Bull. Soc. Géol. France*, *4*, 570–87

—— (1905) 'Réponse aux observations de M. Albert Gaudry à propos de sa Note sur le *Chasmotherium*', *Bull. Soc. Géol. France*, *5*, 139

—— (1908) *Les transformations du monde animal*, Paris

Desmond, A. (1975) *The Hot-blooded Dinosaurs*, London

—— (1982) *Archetypes and Ancestors: Palaeontology in Victorian London, 1850–1875*, London

—— (1984) 'Interpreting the origin of mammals: new approaches to the history of palaeontology', *Zool. J. Linn. Soc.*, *82*, 7–16

Dollo, L. (1893) 'Les lois de l'évolution', *Bull. Soc. belge Géol.*, *7*, 164–6

BIBLIOGRAPHY

—— (1901) 'Sur l'origine de la tortue luth (*Dermochelys coriacea*)', *Bull. Soc. roy. sc. méd. nat. Bruxelles*, 1–26
—— (1923) 'Le centenaire des Iguanodons (1822–1922)', *Phil. Trans. Roy. Soc. London*, B, *212*, 67–78
Dong, Z.M. (1985) 'The study of dinosaur fossils in China' in *Dinosaurs in China*, Hong Kong
Duméril, C. (1838) 'Remarques à l'occasion du mémoire de M. de Blainville (didelphes de Stonesfield)', *C. R. Acad. Sc. Paris*, *7*, 736
Edwards, G. (1756) 'An account of *Lacerta (Crocodilus) ventre marsupio donato, faucibus Mergansensis rostrum aemulantibus*', *Phil. Trans. Roy. Soc. London*, *49*, 2, 639–42
Edwards, W.N. (1976) *The Early History of Palaeontology*, London
Eudes-Deslongchamps, J.A. (1838) 'Mémoire sur le *Poekilopleuron Bucklandii*, grand saurien fossile intermédiaire entre les crocodiles et les lézards', *Mém. Soc. Linn. Normandie*, *6*, 37–146
—— (1896) 'Histoire d'une vocation. Découverte du premier individu de *Teleosaurus cadomensis*', *Bull. Soc. Linn. Normandie*, *10*, 26–49
Fallot, E. (1911) 'Voltaire et la Géologie', *Mém. Soc. Emul. Montbéliard*, *40*, 215–24
Faujas de Saint-Fond, B. (1799) *Histoire naturelle de la montagne de Saint-Pierre de Maestricht*, Paris
Faul, H. and Faul, C. (1983) *It Began With a Stone*, New York
Figuier, L. (1866) *La Terre avant le Déluge* (5th edn), Paris
Filhol, H. (1876) 'Recherches sur les phosphorites du Quercy. Etude des fossiles qu'on y rencontre et spécialement des Mammifères', *Ann. Sc. géol.*, *7*, 7, 1–220
Fischer, H. (1973) 'Johann Jakob Scheuchzer (2. August 1672–23. Juni 1733), Naturforscher und Arzt', *Neujahrsbl. Naturforsch. Ges. Zürich*, 1–168
Flacourt, E. de (1658) *Histoire de la grande isle de Madagascar*, Paris
Flammarion, C. (1886) *Le monde avant la création de l'homme*, Paris
Flaubert, G. (1884) *Bouvard et Pécuchet*, Paris
Forsyth Major, C.I. (1891) 'Le gisement ossifère de Mytilini' in C. de Stefani, C.I. Forsyth Major and W. Barbey, *Samos. Etude géologique, paléontologique et botanique*, Lausanne, 85–99
—— (1894) 'On *Megaladapis madagascariensis*, an extinct gigantic lemuroid from Madagascar, with remarks on the associated fauna and on its geological age', *Phil. Trans. Roy. Soc. London*, B, *185*, 15–38
Fortelius, M. and Kurten, B. (1979) 'Holy Dacian nothing but Roman Grenadier', *Soc. Vert. Paleont. News Bull.*, *116*, 71
Fortis, A. (1802) *Mémoires pour servir à l'histoire naturelle et principalement à l'oryctographie de l'Italie et des pays adjacens*, 2 vols, Paris
Furon, R. (1951) *La Paléontologie*, Paris
Garriga, J. (1796) *Descripcion del esqueleto de un quadrupedo muy corpulento y raro, que se conserva en el Real Gabinete de Historia Natural de Madrid*, Madrid
Gascar, P. (1983) *Buffon*, Paris
Gaudant, J. (1980) 'Louis Agassiz (1807–1873), fondateur de la

205

paléoichthyologie', *Rev. Hist. Sc.*, *33*, 151–62

Gaudry, A. (1862–7) *Animaux fossiles et géologie de l'Attique*, Paris
—— (1866) *Considérations générales sur les animaux fossiles de Pikermi*, Paris
—— (1872) 'Sur des ossements d'animaux quaternaires que M. l'abbé David a recueillis en Chine', *Bull. Soc. Géol. France*, *29*, 177–9
—— (1888) *Les ancêtres de nos animaux dans les temps géologiques*, Paris
—— (1878–90) *Les enchaînements du monde animal dans les temps géologiques*, 3 vols, Paris
—— (1896) *Essai de paléontologie philosophique*, Paris
—— (1905) 'Observations au sujet d'une note sur le *Chasmotherium*', *Bull. Soc. Géol. France*, *5*, 76

Geoffroy Saint-Hilaire, E. (1825) 'Recherches sur l'organisation des gavials', *Mém. Mus. Hist. Nat. Paris*, *12*, 97–155
—— (1831) *Recherches sur de grands sauriens trouvés à l'état fossile vers les confins maritimes de la Basse Normandie, attribués d'abord au crocodile, puis déterminés sous les noms de Téléosaurus et Sténéosaurus*, Paris
—— (1833) 'Palaeontographie. Considérations sur des ossemens fossiles la plupart inconnus, trouvés et observés dans les bassins de l'Auvergne', *Rev. encycl.*, *59*, 76–95
—— (1838) 'De quelques contemporains des Crocodiliens fossiles des âges antédiluviens, d'un rang classique jusque alors indéterminé', *C. R. Acad. Sc. Paris*, *7*, 629–33

Gesner, C. (1558) *De Rerum fossilium, Lapidum et Gemmarum maxime, figuris et similitudinibus Liber*, Zürich

Ginsburg, L. (1984) 'Nouvelles lumières sur les ossements fossiles autrefois attribués au géant Theutobochus', *Ann. Paléont.*, *70*, 3, 181–219

Gosse, I. de (1847) *Histoire naturelle drôlatique et philosophique des Professeurs du Jardin des Plantes*, Paris

Granger, W. (1938) 'Medicine bones', *Natural History*, *42*, 264–71

Grant, R. (1834) 'On the osteology of Rodentia and Marsupialia', *The Lancet*, *2*, 65–73

Gregory, J.T. (1979) 'North American vertebrate paleontology, 1776–1976' in C.J. Schneer (ed.), *Two Hundred Years of Geology in America*, Hanover, N.H., 305–35

Guettard, J.E. (1756) 'Suite du mémoire dans lequel on compare le Canada à la Suisse, par rapport à ses minéraux. Seconde partie. Description des minéraux de la Suisse', *Mém. Acad. Roy. Sc. Paris* (1752), 323–60
—— (1768) *Mémoires sur différentes parties des sciences et des arts*, *1*, Paris
—— (1783) *Mémoires sur différentes parties des sciences et des arts*, *7*, Paris

Hain, J.P. (1672) 'De draconibus Carpathicis', *Misc. curiosa medicophysica acad. nat. curios.*, *3*, 139, 220–58

Halstead, L.B. (1970) '*Scrotum humanun* Brookes 1763 — the first named dinosaur', *J. Insignificant Research*, *5*, 14–15

Harlan, R. (1841) 'A letter from Dr Harlan, addressed to the President, on the discovery of the remains of the *Basilosaurus* or *Zeuglodon*', *Trans. Geol. Soc. London*, 6, 67–8

Hauff, B. (1953) *Das Holzmadenbuch*, Ohringen

Haug, E. (1905) 'Paléontologie' in *Documents scientifiques de la Mission saharienne (Mission Foureau-Lamy)*, Paris, 751–832

Hawkins, T. (1834) *Memoirs of Ichthyosauri and Plesiosauri, Extinct Monsters of the Ancient Earth*, London

—— (1840) *The Book of the Great Sea-dragons, Ichthyosauri and Plesiosauri, Gedolim Taninim of Moses*, London

Hennig, (1912) *Am Tendaguru*, Stuttgart

Heuvelmans, B. (1955) *Sur la piste des bêtes ignorées*, Paris

Hitchcock, E. (1858) *Ichnology of New England. A Report on the Sandstone of the Connecticut Valley, Especially Its Fossil Footmarks. Made to the Government of the Commonwealth of Massachusetts*, Boston

Hoch, E. (1984) 'The influence of Georges Cuvier on the Danish naturalist Peter Wilhelm Lund, "father of Brazilian palaeontology"' in E. Buffetaut, J.M. Mazin and E. Salmon (eds), *Actes du Symposium paléontologique G. Cuvier*, Montbéliard, 273–87

—— (1985) 'Sténon et la Géologie', *Géochronique*, 15, 21–2

Hoffstetter, R. (1959) 'Les rôles respectifs de Brú, Cuvier et Garriga dans les premières études concernant *Megatherium*', *Bull. Mus. Hist. Nat. Paris*, 31, 6, 536–45

Hölder, H. (1960) *Geologie und Paläontologie in Texten und ihrer Geschichte*, Freiburg and Munich

Home, E. (1814) 'Some account of the fossil remains of an animal more nearly allied to fishes than to any of the other classes of animals', *Phil. Trans. Roy. Soc. London*, 104, 571–7

—— (1818) 'Additional facts respecting the fossil remains of an animal, on the subject of which two papers have been printed in the Philosophical Transactions, showing that the bones of the sternum resemble those of the *Ornithorhynchus paradoxus*', *Phil. Trans. Roy. Soc. London*, 108, 24–32

—— (1819) 'An account of the fossil skeleton of the *Proteosaurus*', *Phil. Trans. Roy. Soc. London*, 109, 209–11

Howard, R.W. (1975) *The Dawnseekers*, New York and London

Howe, S.R., Sharpe, T. and Torrens, H.S. (1981) *Ichthyosaurs: a History of Fossil 'Sea-dragons'*, Cardiff

Humboldt, A. von (1835) 'Note sur des empreintes de pieds d'un quadrupède, dans la formation de grès bigarré de Hildburghausen, en Allemagne', *C. R. Acad. Sc. Paris*, 45–8

Hunter, W. (1769) 'Observations on the bones, commonly supposed to be elephant's bones, which have been found near the river Ohio, in America', *Phil. Trans. Roy. Soc. London*, 58, 34–45

Hutchinson, H.N. (1894) *Creatures of Other Days*, London

—— (1897) *Extinct Monsters* (5th edn), London

Huxley, T.H. (1875) 'On *Stagonolepis robertsonii*, and on the evolution of the Crocodilia', *Q. Jl Geol. Soc. London*, 31, 423–38

Ingenieros, J. (1957) *Las doctrinas de Ameghino*, Buenos Aires

Jaeger, G.F. (1824) *De ichthyosauri sive proteosauri fossilis speciminibus in agro bollensi in Wirttembergia repertis*, Stuttgart

Jahn, M.E. (1969) 'Some notes on Dr Scheuchzer and on *Homo diluvii testis*' in C.J. Schneer (ed.), *Toward a History of Geology*, Cambridge and London

Janensch, W. (1914) 'Bericht über den Verlauf der Tendaguru-Expedition', *Arch. Biontol.*, *3*, 17–58

Jefferson, T. (1799) 'A memoir on the discovery of certain bones of a quadruped of the clawed kind in the western parts of Virginia', *Trans. Amer. Phil. Soc.*, *4*, 246–60

Kaup, J.J. (1835) 'Thier-Fährten von Hildburghausen; *Chirotherium* oder *Chirosaurus*', *N. Jb. Mineralogie*, 327–8

Kellogg, R. (1936) *A Review of the Archaeoceti*, Washington

Kirby, W. (1835) *On the Power, Wisdom and Goodness of God as Manifested in the Creation of Animals*, London

Kircher, A. (1678) *Mundus subterraneus*, Amsterdam

Knorr, G.W. and Walch, J.E. (1755–78) *Sammlung der Merkwürdigkeiten der Natur, und Altertümer des Erdbodens, welche petrificierte Körper enthält*, Nuremberg

Koken, E. (1885) 'Über fossile Säugethiere aus China, nach den Sammlungen des Herrn Ferdinand Freiherrn von Richthofen bearbeitet', *Geol. Pal. Abh.*, *3*, 31–113

Kovalevskii, V. (1873a) 'Sur l'*Anchitherium aurelianense* Cuv. et sur l'histoire paléontologique des chevaux', *Mém. Acad. Sc. St Petersburg*, *20*, 1–73

—— (1873b) 'Monographie der Gattung *Anthracotherium*', *Palaeontographica*, *22*, 131–346

Krebs, B. (1966) 'Zur Deutung der *Chirotherium*-Fährten', *Natur und Museum*, *96*, 389–96

Lamarck, J.B. (1809) *Philosophie zoologique*, Paris

Lambe, L. (1905) 'The progress of vertebrate palaeontology in Canada', *Proc. Trans. Roy. Soc. Canada*, 2, *10*, 1, 13–56

Laming-Emperaire, A. (1964) *Origines de l'archéologie préhistorique en France*, Paris

Langenmantel, H.A. (1688) 'De ossibus elephantum', *Misc. Curiosa*, 7, 446–7

Lanham, U. (1973) *The Bone Hunters*, New York and London

Lartet, E. (1851) *Notice sur la colline de Sansan*, Auch

Laurent, G. (1984) 'Cuvier et le catastrophisme' in E. Buffetaut, J.M. Mazin and E. Salmon, *Actes du Symposium Paléontologique G. Cuvier*, Montbéliard, 337–46

Lavauden, L. (1931) 'Animaux disparus et légendaires de Madagascar', *Rev. scient. illustr.*, *10*, 297–308

Leeds, E.T. (1956) *The Leeds Collection of Fossil Reptiles from the Oxford Clay of Peterborough*, Oxford

Le Gros Clark, W.E. and Leakey, L.S.B. (1951) 'The Miocene Hominoidea of East Africa', *Fossil Mammals of Africa*, *1*, 1–117

Leibniz, G.W. (1749) *Protogaea, sive de prima facie telluris et antiquissimae historiae vestigiis in ipsis naturae monumentis dissertatio*, Göttingen

Leidy, J. (1853) 'The ancient fauna of Nebraska: or, a description of remains of extinct Mammalia and Chelonia, from the Mauvaises Terres of Nebraska', *Smithsonian Contrib. Knowl.*, 6, 1–126

—— (1869) *The Extinct Mammalian Fauna of Dakota and Nebraska, Including an Account of Some Allied Forms from Other Localities, Together With a Synopsis of the Mammalian Remains of North America*, Philadelphia

Le Mesle, G. and Péron, P.A. (1880) 'Sur des empreintes de pas d'oiseaux observées par M. Le Mesle dans le Sud de l'Algérie', *C. R. Ass. Franc. Avanc. Sc.*, 9, 528–33

Leroi-Gourhan, A. (1955) *Les hommes de la Préhistoire. Les chasseurs*, Paris

Lhwyd, E. (1699) *Eduardi Luidii apud oxonienses cimeliarchae ashmoleani lithophylacii britannici ichnographia*, London

Linck, H. (1718) 'Excerpta ex litteris Henrici Linckii ad V.Cl.J. Woodwardum', *Acta eruditorum*, 188–9

Link, H.F. (1835) 'Note sur les traces de pattes d'animaux inconnus trouvées près de Hildburghausen en Saxe', *C. R. Acad. Sc. Paris*, 258–60

Lorenz von Liburnau, L. (1901) 'Uber einige Reste ausgestorbener Primaten von Madagaskar', *Denkschr. kaiserl. Akad. Wiss., math. nat. Cl.*, 70, 1–15

Lydekker, R. (1888–90) *Catalogue of the Fossil Reptilia and Amphibia in the British Museum (Natural History)*, London

—— (1893) 'Contributions to a study of the fossil vertebrates of Argentina. I.', *An. Mus. La Plata*, 2, 1–91

—— (1895) 'On bones of a sauropodous dinosaur from Madagascar', *Q. Jl Geol. Soc. London*, 51, 329–36

Lyell, C. (1830) *Principles of Geology*, London

—— (1855) *A Manual of Elementary Geology* (5th edn), London

Mc Cartney, P.J. (1977) *Henry De la Beche: Observations on an Observer*, Cardiff

Mantell, G.A. (1825) 'Notice on the *Iguanodon*, a newly discovered fossil reptile, from the sandstone of Tilgate Forest, in Sussex', *Phil. Trans. Roy. Soc. London*, 115, 179–86

—— (1838) *The Wonders of Geology*, London

—— (1846) 'A few notes on the prices of fossils', *London Geol. Jl*, 1, 13–17

—— (1851) *Petrifications and Their Teachings: or a Hand-book to the Gallery of Organic Remains of the British Museum*, London

Marsh, O.C. (1880) 'Odontornithes: a monograph on the extinct toothed birds of North America', *Mem. Peabody Mus. Yale Coll.*, 1, 1–201

—— (1896) 'The dinosaurs of North America', *Ann. Rep. US Geol. Surv.*, 16, 133–244

Martin, G.P.R. (1966) 'Ein "Diskussionsbeitrag" Eduard Mörike's zum *Chirotherium*-Problem', *Natur und Museum*, 96, 1, 9–11

Miller, H. (1841) *The Old Red Sandstone, or New Walks in an Old Field*, Edinburgh

Milne-Edwards, A. and Grandidier, A. (1869) 'Nouvelles observations

sur les caractères zoologiques et les affinités naturelles de l'*Aepyornis* de Madagascar', *Ann. Sc. nat.*, 7, 167–96
—— and —— (1894) 'Observations sur les *Aepyornis* de Madagascar', *C. R. Acad. Sc. Paris, 118*, 122–7

Miquel, J. (1896) 'Note sur la géologie des terrrains secondaires et des terrains tertiaires du département de l'Hérault', *Bull. Soc. Et. Sc. Nat. Béziers*, *19*, 31–74

Molyneux, T. (1697) 'A Discourse concerning the large Horns frequently found under Ground in Ireland, Concluding from them that the great American Deer, call'd a Moose, was formerly common in that Island: With Remarks on some other things Natural to that country', *Phil. Trans. Roy. Soc. London*, *19*, 489–512

Morello, N. (1979) *La nascita della paleontologia nel seicento. Colonna, Stenone e Scilla*, Milano

Moreno, F.P. (1891a) 'El Museo de La Plata. Rapida ojeada sobre su fundacion y desarollo', *Rev. Mus. La Plata*, *1*, 28–55

—— (1891b) 'Reseña general de las adquisiciones y trabajos hechos en 1889 en el Museo de La Plata', *Rev. Mus. La Plata*, 58–70

—— (1979) *Reminiscencias de Francisco P. Moreno*, Buenos Aires

Muir Wood, R. (1985) *The Dark Side of the Earth*, London

Murchison, C. (1868) 'Biographical sketch' in C. Murchison (ed.), *Palaeontological Memoirs and Notes of the Late Hugh Falconer*, *A.M.*, *M.D.*, London, *1*, xxiii–liii

Neumayr, M. (1895) *Erdgeschichte*, Leipzig and Vienna

Newton, R.B. (1893) 'On the discovery of a Secondary reptile in Madagascar: *Steneosaurus Baroni*; with a reference to some post-Tertiary vertebrate remains from the same country recently acquired by the British Museum (Natural History)', *Geol. Mag.*, *10*, 193–8

Nopcsa, F. (1899) 'Dinosaurierreste aus Siebenbürgen (Schädel von *Limnosaurus transsylvanicus* nov. gen. et spec.)', *Denkschr. Kaiserl. Akad. Wiss.*, *math. nat. Kl.*, *68*, 1–37

Nordmann, A. von (1858–60) *Palaeontologie Südrusslands*, Helsinki

Norman, D.B. (1980) 'On the ornithischian dinosaur *Iguanodon bernissartensis* of Bernissart (Belgium)', *Mém. Inst. roy. Sc. nat. Belgique*, *178*, 1–105

—— (1985) *The Illustrated Encyclopedia of Dinosaurs*, London

Oakley, K.P. (1975) 'Decorative and symbolic uses of vertebrate fossils', *Pitt Rivers Mus. Occas. Pap. Technol.*, *12*, 1–60

Omalius d'Halloy, J.J. (1843) *Précis élémentaire de Géologie*, Paris

Orbigny, A. d' (1849–52) *Cours élémentaire de paléontologie et de géologie stratigraphiques*, Paris

Osborn, H.F. (1930) *Fifty-two Years of Research, Observation and Publication, 1877–1929*, New York

Ostrom, J.H. (1985) 'Introduction to *Archaeopteryx*' in M.K. Hecht, J.H. Ostrom, G. Viohl and P. Wellnhofer (eds), *The Beginnings of Birds*, Eichstätt, pp. 9–20

Outram, D. (1984) *Georges Cuvier. Vocation, Science and Authority in Post-revolutionary France*, Manchester

Owen, R. (1838a) 'On the jaws of the *Thylacotherium prevostii* (Valenciennes) from Stonesfield', *Proc. Geol. Soc. London*, *3*, 5–9

—— (1838b) 'Fossil remains from Wellington Valley, Australia' in T. Mitchell, *Three Expeditions into the Interior of Eastern Australia, With Descriptions of the Recently Explored Region of Australia Felix, and of the Present Colony of New South Wales*, London, 2, 359–63

—— (1841) 'Description of parts of the skeleton and teeth of five species of the genus *Labyrinthodon* (*Lab. leptognathus*, *Lab. pachygnathus*, and *Lab. ventricosus*, from the Coton-end and Cubbington Quarries of the Lower Warwick Sandstone; *Lab. Jaegeri*, from Guy's Cliff, Warwick; and *Lab. scutulatus*, from Leamington); with remarks on the probable identity of the *Cheirotherium* with this genus of extinct Batrachians', *Trans. Geol. Soc. London*, 6, 2, 515–43

—— (1842) 'Report on British fossil reptiles', *Rep. Brit. Ass. Adv. Sc.*, 11, 60–204

—— (1843) 'On the discovery of the remains of a mastodontoid pachyderm in Australia', *Ann. Mag. Nat. Hist.*, 11, 7–12

—— (1845) 'Description of certain fossil crania discovered by A.G. Bain, Esq., in the sandstone rocks at the southeastern extremity of Africa, referable to different species of an extinct genus of Reptilia (*Dicynodon*) and indicative of a new tribe or suborder of Sauria', *Trans. Geol. Soc. London*, 2, 7, 59–84

—— (1863) 'On the fossil remains of a long-tailed bird (*Archaeopteryx macrurus*, Ow.) from the lithographic slate of Solenhofen', *Proc. Roy. Soc. London*, 12, 272–3

—— (1870) 'On fossil remains of mammals found in China', *Q. Jl Geol. Soc. London*, 26, 417–34

—— (1878) 'On the influence of the advent of a higher form of life in modifying the structure of an older and lower form', *Q. Jl Geol. Soc. London*, 34, 421–30

Pallas, P.S. (1779) *Observations sur la formation des montagnes et les changements arrivés au globe, pour servir à l'histoire naturelle de M. le comte de Buffon*, Saint Petersburg

Peale, R. (1803) *An Historical Disquisition on the Mammoth, or Great American Incognitum, an Extinct, Immense, Carnivorous Animal Whose Remains Have Been Found in North America*, London

Pentland, J.B. (1828) 'Description of fossil remains of some animals from the North East Border of Bengal', *Trans. Geol. Soc. London*, 2, 393–4

Peters, E. (1930) *Die Altsteinzeitliche Kulturstätte Petersfels*, Augsburg

Pfizenmayer, E.W. (1926) *Mammutleichen und Urwaldmenschen in Nordost-Sibirien*, Leipzig

Phillips, J. (1871) *Geology of Oxford and the Valley of the Thames*, Oxford

Pictet, F.J. (1853) *Traité de Paléontologie*, Geneva

Piveteau, J. (1951) *Images des mondes disparus*, Paris

Platt, J. (1758) 'An Account of the fossile Thigh-bone of a large Animal, dug up at Stonesfield, near Woodstock, in Oxfordshire', *Phil. Trans. Roy. Soc. London*, 50, 524–7

Plot, R. (1676) *The Natural History of Oxford-shire, Being an Essay Toward the Natural History of England*, Oxford

Pouchet, F.A. (1872) *L'Univers*, Paris

Prévost, C. (1824) 'Observations sur les schistes calcaires oolitiques de Stonesfield en Angleterre, dans lesquels ont été trouvés plusieurs ossemens fossiles de Mammifères', *Ann. Sc. Nat.*, 4, 389–417

Prout, H.A. (1847) 'Description of a fossil maxillary bone of a *Palaeotherium*, from near White River', *Silliman's Journal*, 3, 248–50

Riabinin, A.N. (1930) '*Mandschurosaurus amurensis* nov. gen. nov. sp., a hadrosaurian dinosaur from the Upper Cretaceous of Amur River', *Mém. Soc. Paléont. Russie*, 2, 1–37

Robinet, J.B. (1768) *Considérations philosophiques de la gradation naturelle des formes de l'être, ou les essais de la nature qui apprend à faire l'homme*, Paris

Roume, P.R. (1796) 'Squelette fossile trouvé sur les bords de La Plata', *Bull. Soc. Philomath. Paris*, 1, 96–7

Rudwick, M.J.S. (1972) *The Meaning of Fossils*, London

Rupke, N.A. (1983) *The Great Chain of History. William Buckland and the English School of Geology (1814–1849)*, Oxford

Ruse, M. (1979) *The Darwinian Revolution*, Chicago and London

Scheuchzer, J.J. (1708) *Piscium querelae et vindiciae*, Zürich

—— (1726) *Homo diluvii testis et theoskopos*, Zürich

—— (1731) *Physica sacra*, Augsburg and Ulm

Schlosser, M. (1902) 'Die fossilen Säugethiere China's', *Centralbl. Min. Geol. Pal.*, 529–35

Schneider-Hauff, G. (1973) *Saurier tauchen auf. . . Aus dem Leben und der Arbeit von Bernhard Hauff*, Leinfelden

Schwarzbach, M. (1980) *Alfred Wegener und die Drift der Kontinente*, Stuttgart

Scilla, A. (1670) *La Vana Speculazione disingannata dal Senso. Lettera risponsiva circa i corpi marini, che petrificati si trovano in varii luoghi terrestri*, Naples

Seeley, H.G. (1887) 'On the classification of the fossil animals commonly named Dinosauria', *Proc. Roy. Soc. London*, 43, 165–71

—— (1901) *Dragons of the Air*, London

Shor, E.N. (1974) *The Fossil Feud between E.D. Cope and O.C. Marsh*, Hicksville

Sickler, F.K.L. (1835) 'Sendschreiben an J.F. Blumenbach über die höchstmerkwürdigen, vor einigen Monaten erst entdeckten Reliefs der Fährten urweltlicher grosser und unbekannter Thiere in den Hessberger Sandsteinbrüchen bei der Stadt Hildburghausen', *N. Jb. Mineralogie*, 230–4

Simpson, G.G. (1942) 'The beginnings of vertebrate paleontology in North America', *Proc. Am. Phil. Soc.*, 86, 130–88

—— (1944) *Tempo and Mode in Evolution*, New York

—— (1951) *Horses*, New York

—— (1984) *Discoverers of the Lost World*, New Haven and London

Soemmerring, S.T. von (1812) 'Uber einen *Ornithocephalus*', *Denkschr. Kgl. Bayer. Akad. Wiss., math. phys. Cl.*, 3, 89–158

Soergel, W. (1925) *Die Fährten der Chirotheria. Eine paläobiologische Studie*, Jena

Solounias, N. (1981) 'The Turolian fauna from the island of Samos,

Greece', *Contrib. Vert. Evol.*, 6, 1–232

Spener, C.M. (1710) 'Disquisitio de crocodilo in lapide scissili expresso, aliisque Lithozois', *Misc. berolin. Increm. Sc.*, 1, 92–110

Spix, J.B. von and Martius, C.F.P. von (1828) *Reise in Brasilien*, Munich

Stearn, W.T. (1981) *The Natural History Museum at South Kensington*, London

Steno, N. (1667) *Elementorum myologiae specimen, seu musculi descriptio geometrica. Cui accedunt canis carchariae dissectum caput, et dissectis piscis ex canum genere*, Florence

——— (1669) *De solido intra solidum naturaliter contento dissertationis prodromus*, Florence

Stromer, E. (1916) 'Richard Markgraf und seine Bedeutung für die Erforschung der Wirbeltierpaläontologie Ägyptens', *Centralbl. Mineral. Geol.*, 287–8

Swedenborg, E. (1734) *Regnum subterraneum sive minerale de cupro et orichalco*, Dresden and Leipzig

Taquet, P. (1984) 'Cuvier, Buckland, Mantell et les dinosaures' in E. Buffetaut, J.M. Mazin and E. Salmon (eds), *Actes du Symposium Paléontologique G. Cuvier*, Montbéliard, 475–94

Tasnadi Kubacska, A. (1945) *Franz Baron Nopcsa*, Budapest

Tedford, R. (1984) 'The Diprotodons of Callabonna' in M. Archer and G. Clayton (eds), *Vertebrate Zoogeography and Evolution in Australasia*, Carlisle, 999–1002

Tentzel, W.E. (1697) 'Epistola de sceleto elephantino Tonnae nuper effosso', *Phil. Trans. Roy. Soc. London*, 19, 234, 757–76

Thevenin, A. (1910) 'Albert Gaudry. Notice nécrologique', *Bull. Soc. Géol. France*, 10, 351–74

Tobien, H. (1974) 'Meyer, Christian Erich Hermann von' in C.C. Gillispie (ed.), *Dictionary of Scientific Biography*, New York, 9, 345–6

——— (1984) 'Johann Heinrich Merck (1741–1791) und die Wirbeltierpaläontologie des ausgehenden 18. Jahrhunderts' in E. Buffetaut, J.M. Mazin and E. Salmon (eds), *Actes du Symposium paléontologique G. Cuvier*, Montbéliard, 495–512

Todes, D.P. (1978) 'V.O. Kovalevskii: the genesis, content and reception of his paleontological work', *Stud. Hist. Biol.*, 2, 99–165

Torrubia, J. (1754) *Aparato para la historia natural española*, Madrid

Touret, L. (1984) 'Georges Cuvier et le Teylers Museum, au travers des notes et de la correspondance de l'époque' in E. Buffetaut, J.M. Mazin and E. Salmon (eds), *Actes du Symposium Paléontologique G. Cuvier*, Montbéliard, 513–34

Valenciennes, A. (1838) 'Observations sur les mâchoires fossiles des couches oolithiques de Stonesfield, nommées *Didelphis prevostii* et *D. bucklandii*', *C. R. Acad. Sc. Paris*, 7, 572–80

Verne, J. (1864) *Voyage au centre de la Terre*, Paris

Villot, R. (1957) *Auguste Pomel, démocrate et savant*, Oran

Viohl, G. (1985) 'Carl F. and Ernst O. Häberlein, the sellers of the London and Berlin specimens of *Archaeopteryx*' in M.K. Hecht, J.H. Ostrom, G. Viohl and P. Wellnhofer (eds), *The Beginnings of*

Birds, Eichstätt, pp. 349–52

Voigt, F.S. (1835) 'Thier-Fährten im Hildburghauser Sandsteine (*Palaeopithecus*)', *N. Jb. Mineralogie*, 322–6

—— (1836) 'Weitere Nachrichten über die Hessberger Thierfährten', *N. Jb. Mineralogie*, 166–74

Vollgnad, H. (1673) 'De draconibus carpathicis et transsylvanicis', *Misc. curios. Ephemer. Acad. Leop. Carol.*, 4–5, 226–9

Wagner, A. (1839) 'Fossile Ueberreste von einem Affenschädel und anderen Säugethieren aus Griechenland', *Gelehrte Anz. Bayer. Akad. Wiss.*, 8, 306–11

—— (1858) *Geschichte der Urwelt*, Leipzig

—— (1861) 'Uber ein neues, angeblich mit Vogelfedern versehenes Reptil aus dem Solnhofener lithographischen Schiefer', *Sitzungsber. Bayer. Akad. Wiss.*, 2, 146–54

Wegener, A. (1929) *Die Entstehung der Kontinente und Ozeane*, Braunschweig

Weishampel, D.B. (1984) 'Trossingen: E. Fraas, F. von Huene, R. Seemann, and the "Schwäbische Lindwurm" *Plateosaurus*' in W.E. Reif and F.Westphal (eds), *Third Symposium on Mesozoic Terrestrial Ecosystems, Short Papers*, Tübingen, 249–53

—— and Reif, W.E. (1984) 'The work of Franz Baron Nopcsa (1877– 1933): dinosaurs, evolution and theoretical tectonics', *Jb. Geol. Bundesanst.*, 127, 2, 187–203

Welles, S.P. and Gregg, D.R., (1971) 'Late Cretaceous marine reptiles of New Zealand', *Rec. Canterbury Mus.*, 9, 1, 1–111

Wellnhofer, P. (1980) 'The history of the Bavarian State Collection of Palaeontology and Historical Geology in Munich', *J. Soc. Biblphy Nat. Hist.*, 9, 4, 383–90

—— (1983) *Solnhofener Plattenkalk: Urvögel und Flugsaurier*, Maxberg

—— (1984) 'Cuvier and his influence on the interpretation of the first known pterosaur' in E. Buffetaut, J.M. Mazin and E. Salmon (eds), *Actes du Symposium Paléontologique G. Cuvier*, Montbéliard, 535–48

—— (1985) 'The story of Albert Oppel's *Archaeopteryx* drawing' in M.K. Hecht, J.H. Ostrom, G. Viohl and P. Wellnhofer (eds), *The Beginnings of Birds*, Eichstätt, pp. 353–7

Wendt, H. (1971) *Ehe die Sintflut kam*, Munich

Weigmann, A.F.A. (1835) 'Thierfährten im bunter Sandsteine', *Arch. Naturgesch.*, 1, 127–31

Williamson, W.C. (1867) 'On a cheirotherian footprint from the base of the Keuper of Daresbury, Cheshire', *Q. Jl Geol. Soc. London*, 23, 56–7

Woodward, J. (1695) *An Essay Toward a Natural History of the Earth: and Terrestrial Bodies, especially Minerals: as also of the Seas, Rivers and Springs. With an Account of the Universal Deluge: and of the Effects that it had upon the Earth*, London

Woodward, A.S. (1918) 'Vladimir Prochorovitch Amalitsky. Obituary', *Geol. Mag.*, 5, 431–2

Wooller, (1758) 'A description of the fossil skeleton of an animal found in the alum rock near Whitby', *Phil. Trans. Roy. Soc. London, 50*,

786–91
Zittel, K.A. von (1890–3) *Handbuch der Palaeontologie* (vols 3 and 4),
   Munich and Leipzig

Index